Lecture Notes in Mathematics

Edited by A. Dold and B. Eckmann

Series: Forschungsinstitut für Mathematik, ETH Zürich

845

Allen Tannenbaum

Invariance and System Theory: Algebraic and Geometric Aspects

Springer-Verlag
Berlin Heidelberg New York 1981

Author

Allen Tannenbaum
Department of Theoretical Mathematics
The Weizmann Institute of Science
Rehovot/Israel

AMS Subject Classifications (1980): 14D20, 14D25, 93B15, 93B25, 93D15

ISBN 3-540-10565-4 Springer-Verlag Berlin Heidelberg New York
ISBN 0-387-10565-4 Springer-Verlag New York Heidelberg Berlin

Printing and binding: Beltz Offsetdruck, Hemsbach/Bergstr.
2141/3140-543210

Math
sep.

To Rina

TABLE OF CONTENTS

Introduction

Notation and Terminology

Part I.	Some Basic Algebraic Geometry	1
	1. Affine Geometry	1
	2. Projective Geometry	3
	3. Regular Mappings	6
	4. The Constructibility Theorem of Chevalley	9
	5. Sheaves and Vector Bundles	10
	6. The Grassman Variety	14
	7. Some Remarks on Algebraic Geometry Over a Non-Algebraically Closed Field	16
Part II.	Some Basic System Theory	18
	1. Dynamical Systems	18
	2. Controllability and Reachability	22
	3. Constructibility and Observability	24
	4. State Variables	26
	5. Transfer Functions	32
Part III.	Invariant Theory and Orbit Space Problems	36
	1. Algebraic Groups	36
	2. On the Moduli of Endomorphisms	38
	3. Quotients	42
	4. Reductive Groups and Hilbert's 14th Problem	44
	5. Richardson's Criterion	46
Part IV.	Global Moduli of Linear Time-Invariant Dynamical Systems	48
	1. Complete Reachability and Pre-Stability	48
	2. Construction of the Quotient Space of Completely Reachable Pairs	53
	3. Moduli of Linear Time-Invariant Dynamical Systems	54
	4. The Geometric Structure of the Moduli Space	57
	5. The Global Moduli of Completely Reachable Matrix Triples	59
	6. Some Open Problems	63
Part V.	Local Moduli of Linear Time-Invariant Dynamical Systems	65
	1. Versal Deformations of Matrix Pairs	66
	2. The Control Canonical Form	68
	3. On the Construction of Holomorphic Canonical Forms	71
	4. Versal Deformations of Matrix Triples	74
Part VI.	Algebraic Realization Theory	76
	1. Input/Output Maps and Abstract Realization Theory	76
	2. Hankel Matrices	80
	3. Realizations of Rational Matrices	84
	4. Partial Realizations	87
	5. Systems over Rings	89
	6. Polynomial Systems	96
Part VII.	On the Geometry of Rational Transfer Functions	104
	1. Cauchy Indices and the Connected Components of Rat(n)	104
	2. Complex Transfer Functions	108
	3. The Topology of Rat(n)	113
	4. Partial Realizations (Again)	116

Part VIII. Feedback and Stabilization of Systems with Parameter Uncertainty 119

1. Classical Stability Theory 119
2. Kronecker Indices and State Feedback 122
3. Coefficient and Pole Assignability 130
4. Blending and Output Feedback 133
5. Interpolation in the Unit Disc 139
6. Feedback Stabilization of Plants with Uncertainty in the Gain Factor 143

Bibliography 150

Index 158

INTRODUCTION

These lecture notes are based on a series of lectures given by the author at the Mathematical System Theory Institute of the ETH, Zürich, in the Spring of 1980. The purpose of the lectures were two-fold:

(1) To introduce theoretical mathematicians to some of the ideas and methods of system theory, and to attempt to convince them that despite the applied aspects of the field that there are, in system theory, some interesting and deep problems from pure mathematics to be solved and thus for which their techniques could be of great value.

(2) To introduce people working in system theory to the ideas of algebraic geometry, differential geometry, algebraic topology, and invariant theory (but with by far the strongest emphasis on algebraic geometry and geometric invariant theory) which are currently becoming popular in algebraic system theory.

The choice of topics given here was of course dictated by the author's personal tastes and prejudices, and also by the research interests of those working at the ETH in system theory during 1980. Thus even though we believe we have covered a suitably wide range of topics, we make no claims of completeness. As a supplement to these notes, there are now two very nice survey papers in the literature about applications of algebraic geometry in system theory by Byrnes-Falb [19] and Hazewinkel [56].

We now briefly sketch the contents of these notes. Part I is concerned with giving just that part of algebraic geometry which we feel to be most useful in system theory. It should be regarded more as a guide to the literature than anything else, but we hope it will at least acquaint the reader with some of the more important concepts of algebraic geometry.

In Part II, there is a similar sketch of the basic notions of system and control theory. An added feature however in our treatment is an argument due to Deligne [26] concerning the equivalence of the state space and differential equation formulation of certain kinds of systems.

In Part III, we discuss those facts about algebraic groups and geometric invariant theory that we will need in Part IV. In particular, we discuss some basic moduli problems (based on Mumford [104] and [108]) as well as defining the important notions of "quotient" and "geometric quotient".

Perhaps the first non-trivial use of algebraic geometry (and implicitly geometric invariant theory) in system theory is due to R. Kalman who described the quotient space of completely reachable systems of fixed dimension modulo the action of the general linear group as a smooth quasi-projective variety. To do this he essentially generalized the Grassmannian construction. (See Kalman [79] for details as well as Part IV, Section 2 of these notes.) This construction has had many important consequences among which showing that except in the scalar input case, there exist no global algebraic canonical forms (Hazewinkel-Kalman [58] and Part IV, Section 4 of these notes). There has since been quite a large literature about the Kalman construction and its generalizations, e.g., Hazewinkel [54], Byrnes-Gauger [20], Byrnes-Hurt [21]. We will try in Part IV to present these ideas in as simple a form as possible, and specifically as natural constructions coming from Mumford's geometric invariant theory as described in Part III. In point of fact, we will see that the moduli properties (Hazewinkel [53], [54]) and even the existence of the quotient space of completely reachable systems as a smooth quasi-projective variety, should be regarded as generalizations of Mumford's treatment of endomorphisms as given in his Oslo talks [108].

We have remarked already that except in the scalar input case, there exist no global algebraic canonical forms. This still leaves open the question of local canonical forms. Generalizing some very nice work of Arnold [5], in Part V of these notes we will see how to derive local canonical forms not only for completely reachable systems, but also for arbitrary linear time-invariant dynamical systems.

In Part VI we describe the so-called "system realization problem" which concerns the construction of a state space model of a system from its input/output behavior. This will give us an opportunity to describe some of the results about linear systems defined over commutative rings in which a number of techniques from algebraic geometry will be needed. Moreover, we will discuss some recent work of Sontag [139] about realizations of polynomial input/output maps which will give us a first-hand look at some of the properties of dynamical systems which are not necessarily linear.

Part VII is concerned with the geometry of rational transfer functions and since here some algebraic topology is used (the necessary definitions are included), the flavor is of a different type than the preceding parts. We have included this topic since besides its importance in system identification theory, we will see that the results indeed relate to some of the preceding work as well as the blending problem described in Part VIII.

Finally in Part VIII, we come to a topic of great classical and modern importance, namely stabilization through feedback. We describe in some detail the results about state feedback including the famous "pole-shifting theorem" as well as more recent results concerning systems defined over rings. We discuss the work of Youla et al [154] concerning stabilization through dynamic output feedback, and the generalization of this work in what is known as the "blending problem". We include at the end a specific system design.

We would like to thank Professor Rudolf Kalman for inviting us to give these lectures as well as introducing us to the whole subject of algebraic system theory. We also benefitted a great deal from numerous conversations with M. Hazewinkel about algebro-geometric methods in system theory, and H. Kraft about invariant theory (especially the applications in Part IV).

This work was done while the author was a guest at the Forschungsinstitut of the ETH, Zürich. The author spent a splendid two years at this institution, and he wishes to sincerely thank Professor Beno Eckmann for inviting him to come and for his kindness during the author's stay. We also would like to thank the staff for their hospitality. Finally, for the superb job of typing this manuscript, we thank Mrs. Ruby Musrie of the Weizmann Institute of Science, Rehovot, Israel.

NOTATION AND TERMINOLOGY

We have tried to follow the standard notational and terminological practices common to pure mathematics and algebraic system theory. We list here a few of the more commonly used symbols, conventions, and prerequisites for reading the text:

(i) If G is a group acting on a set S, the stabilizer subgroup of $s \in S$ will be denoted by $\mathrm{stab}(s)$, and the orbit of s by $O(s)$.

(ii) For R a commutative ring with unity, $M_{n,m}(R)$ will stand for the set of $n \times m$ matrices with coefficients in R.

(iii) If $A \in M_{n,m}(R)$, then $A^t \in M_{m,n}(R)$ will stand for the transpose of A.

(iv) For R a commutative ring with unity, we use the symbols $GL(n,R)$ and $GL(R^n)$ interchangeably for the group of R-linear automorphisms of R^n.

(v) $\mathbb{R}$: = field of real numbers.

(vi) $\mathbb{C}$: = field of complex numbers.

(vii) $\mathbb{N}$: = set of positive integers.

(viii) We will use several times in the text some standard terms from category theory (for example "contravariant functor", "morphism of functors", etc.). For accounts of all this see [91] or [142].

(ix) The basic definitions needed from algebraic geometry and system theory will of course be given in the text. In certain places we will use some elementary notions from the theory of complex spaces and complex manifolds. Usually we will define all the relevant terms. However for some of the standard terminology which may not be familiar to the reader, everything we use may be found in references [24], [45], [49], [103] and [151] (as well as other places).

PART I. SOME BASIC ALGEBRAIC GEOMETRY

In this part we would like to sketch those results of algebraic geometry which are most immediately applicable to algebraic system theory and which will be essential for us in these lectures. No attempt at completeness will be made and several key results will be left as exercises or references given.

There are several good books in print now on the topics we will cover here. We recommend however the books of Griffiths-Harris [45], Hartshorne [52], and Mumford [107] as being especially good. Our treatment here is in point of fact a conglomeration of topics from all three of the above references.

§1. Affine Geometry

Throughout Part I, k will denote a fixed algebraically closed field of arbitrary characteristic.

Let $R := k[X_1,\ldots,X_n]$ be the polynomial ring over k in n variables. Then elements $f \in R$ define functions on k^n in the obvious way. Namely, for $p = (a_1,\ldots,a_n) \in k^n$, $f(p) := f(a_1,\ldots,a_n)$. Let $T \subset R$ be any subset. Define $V(T) := \{p \in k^n \mid f(p) = 0 \text{ for all } f \in T\}$. Clearly if $\mathcal{A}$ = ideal generated by the elements of T, $V(\mathcal{A}) = V(T)$. Moreover since R is Noetherian $\mathcal{A}$ is generated by finitely many elements, say $f_1,\ldots,f_m$, and therefore $V(T) = \{p \in k^n \mid f_1(p) = \ldots = f_m(p) = 0\}$. Consequently we make the following definition:

Definition (1.1). An affine algebraic variety is a subset of k^n of the form $V(T)$ for some $T \subset R$.

Proposition (1.2). *The intersection of any number of algebraic varieties is a variety. The union of a finite number of varieties is a variety. The null set and k^n are varieties.*

Proof. We leave this as a nice exercise or see e.g. Hartshorne [52], page 2.

Q.E.D.

Via (1.2) we can make the following definition:

Definition (1.3). The Zariski topology on k^n is the topology defined by letting the closed subsets be the affine varieties of k^n.

Notation (1.4). $\mathbb{A}^n$ stands for k^n with the Zariski topology.

Example (1.5). Consider $\mathbb{A}^1$. Then since every polynomial has only finitely many zeroes, closed subsets of $\mathbb{A}^1$ are precisely finite sets of points, the null set, and $\mathbb{A}^1$ itself. In particular $\mathbb{A}^1$ is not Hausdorff.

One can show that $\mathbb{A}^n$ for $n > 0$ is never Hausdorff. We leave this as an exercise or see Hartshorne [52].

Exercise (1.6). Show that the Zariski topology on $\mathbb{A}^2$ is not the product topology on $\mathbb{A}^1 \times \mathbb{A}^1$.

Definition (1.7). A topological space X is said to be irreducible if X cannot be written as a union of two proper closed subsets.

Remark (1.8.). A closed subset of $\mathbb{A}^n$ is many times referred to in the literature as an "algebraic set", and "variety" then in this context means "irreducible algebraic set". We prefer here however to use "variety" in the sense of (1.1).

Definition (1.9). A quasi-affine variety is an open subset of a variety.

Exercise (1.10). (At this point very hard). Give an example of a quasi-affine variety which is not affine.

Via the construction $T \to V(T)$ we have associated varieties to subsets of the polynomial ring R. We now want to go backwards. This is of course very simple. Given $Z \subset \mathbb{A}^n$ any subset, let $I(Z) := \{f \in R \mid f(p) = 0 \text{ for every } p \in Z\}$. $I(Z)$ is called the ideal of Z.

Now the functors V and I are closely related as the following theorem essentially due to Hilbert will show:

Theorem (1.11).

(i) For $T_1 \subseteq T_2 \subseteq R$, $V(T_1) \supseteq V(T_2)$.

(ii) For $Y_1 \subseteq Y_2 \subseteq \mathbb{A}^n$, $I(Y_1) \supseteq I(Y_2)$.

(iii) If $\mathcal{A} \subseteq R$ is an ideal, $I(V(\mathcal{A})) = \sqrt{\mathcal{A}} := \{f \in R \mid f^r \in \mathcal{A} \text{ for some } r > 0\}$. ($\sqrt{\mathcal{A}}$ is called the radical of $\mathcal{A}$.)

(iv) $V(I(Y)) = \bar{Y}$.

In short, the functors I and V set up a 1-1 inclusion reversing correspondence between varieties and ideals $\mathcal{A}$ such that $\mathcal{A} = \sqrt{\mathcal{A}}$. A variety is irreducible if and only if its associated ideal is prime.

Proof. (i), (ii), (iv) are trivial. See e.g. [52], page 3. Part (iii) is not trivial and is precisely the famous Hilbert Nullstellensatz. For a proof of this see e.g. Lang [91], page 256.

Q.E.D.

Theorem (1.11) is very important in that it reduces affine algebraic geometry to ideal theory. It moreover contains a hint of Grothendieck's theory of schemes as

Example (1.12) (iii) will indicate below.

Examples (1.12).

(i) First note that $V(0) = \mathbb{A}^n$. Since (0) is a prime ideal in R, $\mathbb{A}^n$ is an irreducible variety.

(ii) Under the correspondence of (1.11) maximal ideals of R correspond to minimal varieties i.e. to points. Indeed $(a_1,\ldots,a_n)$ will correspond to $(X_1 - a_1,\ldots,X_n - a_n)$ (which gives the form of every maximal ideal of R). This means that to study the geometry of $\mathbb{A}^n$ one can forget about points and just study the maximal ideals of R.

There is of course nothing special about R from this point of view and so in general, given any commutative ring A with identity, let

$$\text{Spec max } A := \{\text{maximal ideals of } A\} .$$

Grothendieck's idea is to instead of looking at just maximal ideals to look at all prime ideals, i.e., he defines

$$\text{Spec } A := \{\text{prime ideals of } A\} .$$

Part of the motivation for doing this is as follows. We note that from (1.11) varieties correspond to radical ideals and so to study varieties we are reduced to studying radical ideals. Now in Noetherian rings like R, every radical ideal may be expressed as an intersection of prime ideals (this is a special case of primary decomposition; for details see Lang [91]), say

$$\mathcal{A} = p_1 \cap \ldots \cap p_s$$

where the p_i are prime. Note that then $V(\mathcal{A}) = V(p_1) \cup \ldots \cup V(p_s)$ and again from (1.11) the $V(p_i)$ are irreducible. Therefore the prime ideals or equivalently the irreducible varieties are the "building blocks" of affine algebraic geometry. In particular to look at Spec R is to look at all the subvarieities of $\mathbb{A}^n$ at once. This approach has turned out to be the most fruitful one in algebraic geometry.

Spec R is a special case of what Grothendieck calls a "scheme". We will not look at this theory any further here but content ourselves with the classical theory of varieties. For elementary scheme theory see Hartshorne [52] or Mumford [106].

§2. Projective Geometry

As is well-known to study the behavior of the roots of polynomial equations at infinity one homogenizes the equations and this leads to the notion of projective space.

More specifically, define projective n-space

$$\mathbb{P}^n : \underset{=\!=}{\text{point set}} \ \mathbb{A}^{n+1} - \{0\} / \sim$$

where $(a_0,\dots,a_n) \sim (a_0',\dots,a_n')$ if and only if there exists $\lambda \in k^*$ such that $a_i = \lambda a_i'$, $i = 0,\dots,n$. Thus as a point set $\mathbb{P}^n$ is just the set of lines through the origin in $\mathbb{A}^{n+1}$. Given a point $p \in \mathbb{P}^n$ we can choose an (n+1)-tuple $(a_0,\dots,a_n)$ which represents the equivalence class defined by p, and the a_i of this (n+1)-tuple are called homogeneous coordinates of p.

What we want to do now is to define on $\mathbb{P}^n$ a topology analogous to the affine Zariski topology of Section 1. To do this we must discuss some elementary properties of graded rings. For a complete discussion of this topic as well as proofs for everything we are about to say, see Zariski-Samuel [156], Vol. II, pages 149-160.

Definitions (2.1).

(i) A graded ring S is a commutative ring with identity, such that S admits a decomposition into a direct sum of abelian groups $\bigoplus_{d\geq 0} S_d$ with the property that $S_d \cdot S_{d'} \subseteq S_{d+d'}$ for all d,d'.

(ii) The elements of S_d are said to be homogeneous of degree d.

(iii) A homogeneous ideal $\mathcal{A} \subseteq S$ is an ideal which enjoys the following equivalent properties:

(a) Given $f \in \mathcal{A}$, all the homogeneous components of f also lie in $\mathcal{A}$;

(b) $\mathcal{A}$ is generated by homogeneous elements.

Example (2.2). The ring $R = k[X_0,\dots,X_n]$ is homogeneous. (Note R in this section has $n+1$ variables.)

Remark (2.3). Let f be homogeneous of degree d. Then the property of f being zero at $(a_0,\dots,a_n)$ depends only on the equivalence class of $(a_0,\dots,a_n)$ in $\mathbb{P}^n$. Thus the expression $f(p) = 0$ makes sense for $p \in \mathbb{P}^n$. In particular if $T \subseteq R$ is a subset consisting of homogeneous elements, we can define

$$V(T) := \{p \in \mathbb{P}^n \mid f(p) = 0 \text{ for all } f \in T\} .$$

For $\mathcal{A}$ a homogeneous ideal,

$$V(\mathcal{A}) := V(\mathcal{A}_T)$$

where $\mathcal{A}_T :=$ {all homogeneous elements of $\mathcal{A}$}. Again by definition of homogeneous ideal and the Noetherian property of R, we can find a finite set of homogeneous generators of $\mathcal{A}$ and hence $V(\mathcal{A})$ is just the locus of zeroes of a finite number of homogeneous polynomials.

Definitions (2.4).

(i) A projective variety is a subset of $\mathbb{P}^n$ of the form $V(T)$ as in (2.3).

(ii) A quasi-projective variety is an open subset of a projective variety.

Proposition (2.5). The intersection of any number of projective varieties is a projective variety, the union of a finite number of projective varieties is a projective variety, the null set and $\mathbb{P}^n$ are projective varieties.

Proof. Exercise or see [52].

Q.E.D.

Definition (2.6). The Zariski topology on $\mathbb{P}^n$ is defined by letting the closed subsets be the projective varieties in $\mathbb{P}^n$.

As in the affine case given $Z \subset P^n$, define

$I(Z) = \{f \in R \text{ homogeneous} \mid f(p) = 0 \text{ for all } p \in Z\}$.

Then we have the following homogeneous analogue of (1.11) a proof of which may be found in Zariski-Samuel [156], Vol. II, pages 171-172 (this is the so-called "homogeneous Nullstellensatz"):

Theorem (2.7). The functors I and V set up a 1-1 inclusion reversing correspondence between projective varieties and homogeneous ideals $\mathfrak{A}$ such that $\mathfrak{A} = \sqrt{\mathfrak{A}}$ with the exception that the ideal $(X_0,\ldots,X_n)$ is not included in this correspondence. Moreover a projective variety is irreducible if and only if its associated ideal is prime.

Remark (2.8). We now come to a rather important point. Namely, that projective varieties admit open affine covers. This fact leads to the notion of "abstract variety" (see e.g. Serre [125]) a notion which we will not explore here.

The idea is to define for each $i = 0,\ldots,n$, $U_i := \{p \in \mathbb{P}^n \mid$ if p is represented by the homogeneous coordinates $(a_0,\ldots,a_n)$, then $a_i \neq 0\}$.

Note then we have bijections of sets $\varphi_i : U_i \to k^n$ defined by $p \longrightarrow (a_0/a_i,\ldots,a_n/a_i)$ with $a_i/a_i = 1$ left out, when p is represented by $(a_0,\ldots,a_n)$. Since we are taking ratios this definition is independent of choice of homogeneous coordinates.

Now the fundamental result is this:

Theorem (2.9). The maps $\varphi_i : U_i \to k^n$ define homeomorphisms from U_i with the induced topology as open subsets of $\mathbb{P}^n$ and k^n regarded as $\mathbb{A}^n$ (i.e., k^n with the Zariski topology).

Proof. Exercise or see [52], pages 10-11.

Q.E.D.

Corollary (2.10). Let $Y \subseteq \mathbb{P}^n$ be a projective variety. Then Y admits an open affine cover. Let $Y' \subseteq \mathbb{P}^n$ be a quasi-projective variety. Then Y' admits an open covering of quasi-affine subspaces.

Proof. Y admits the cover $\{Y \cap U_i\}$ and Y' the cover $\{Y' \cap U_i\}$. Now use (2.9).

Q.E.D.

§3. Regular Mappings

We now have defined the basic objects of algebraic geometry (namely varieties) and so to complete the picture we need to define a proper notion of mapping. It is clear that our mappings should be locally defined in term s of polynomials. The exact definitions are:

Definition (3.1). Let X be a quasi-affine variety (which includes the case of X being affine). Then a map $f : X \to k$ is said to be regular if for every $p \in X$, there exists a neighborhood U of p such that $f|U = g/h$ where g,h are polynomials and h is nowhere vanishing on U.

Exercise (3.2). Show if k is given the Zariski topology (i.e., k is considered to be $\mathbb{A}^1$), then a regular map is continuous.

Given (3.1) the definition for "regularity" in the projective case must be:

Definition (3.3). Let X be a quasi-projective variety (which includes the case of X being projective). Then a map $f : X \to k$ is said to be regular if for every $p \in X$, there exists a neighborhood U of p such that $f|U = g/h$ where g,h are homogeneous polynomials of the same degree and h is nowhere vanishing on U.

Exercise (3.4). Same as (3.2) for the projective case.

We can now define a notion of morphism so that we get a "category" of varieties:

Definition (3.5). Let X,Y be any varieties (quasi-affine or quasi-projective). Then a morphism $\varphi: X \to Y$ is a continuous map such that for every open subset $U \subseteq Y$, and every regular function f on U, the function $f \circ \varphi$ is regular on $\varphi^{-1}(U)$.

Remark (3.6). Classically a morphism was called a "regular mapping" and this terminology is still sometimes used in the literature.

Now so far given the treatments of affine and projective varieties of Sections 1 and 2, there does not seem to be a very big difference other than projective varieties are defined only by homogeneous polynomials. In point of fact, there is a tremendous difference basically because $\mathbb{P}^n$ is complete (an algebraic analogue of compact) while $\mathbb{A}^n$ is not. This fact is reflected in the rings of regular functions:

Definitions (3.7).

(i) Let Y be any quasi-affine or any quasi-projective variety. Then we let $\mathcal{O}(Y) :=$ ring of regular functions on Y.

(ii) Let Y be an affine variety, say $Y \subseteq \mathbb{A}^n$ as a Zariksi closed subset. Let $I(Y)$ be the ideal of Y. Then the coordinate ring $A(Y)$ is defined to be $k[X_1,\ldots,X_n]/I(Y)$.

(iii) Let Y be a projective variety, say $Y \subseteq \mathbb{P}^n$ as a Zariski closed subset. Let $I(Y)$ be the homogeneous ideal of Y. Then the homogeneous coordinate ring $R(Y)$ is defined to be $k[X_0,\ldots,X_n]/I(Y)$.

We now want to understand the precise connections among the definitions (i), (ii), and (iii). For example, for any variety Y, $\mathcal{O}(Y)$ is an invariant of the isomorphism class of Y, i.e. if $Y \cong Y'$, then $\mathcal{O}(Y) \cong \mathcal{O}(Y')$. A priori (ii) and (iii) appear dependent on the embedding of the given variety into some affine or projective space. In point of fact (iii) is dependent on the embedding as the following example will reveal:

Example (3.8). Let $\mathbb{P}^1$ have homogeneous coordinate ring $k[X,Y]$, $\mathbb{P}^2$ have homogeneous coordinate ring $k[U,V,W]$. Represent a point $p \in \mathbb{P}^1$ with homogeneous coordinates (x,y). Then we define a morphism $\varphi\colon \mathbb{P}^1 \to \mathbb{P}^2$ by $\varphi(x,y) = (x^2,xy,y^2)$.

Now one can easily check that φ is an embedding, i.e. $\mathbb{P}^1 \cong \varphi(\mathbb{P}^1)$. Let us see what the homogeneous coordinate ring of $\varphi(\mathbb{P}^1) \subseteq \mathbb{P}^2$ is however. Clearly from the definition of φ, $\varphi(\mathbb{P}^1)$ is defined by the quadratic equation $V^2 - UW = 0$, i.e., $\varphi(\mathbb{P}^1) \subseteq \mathbb{P}^2$ is a conic. (Just note that $(xy)^2 - x^2y^2 = 0$.) Hence $R(\varphi(\mathbb{P}^1)) = k[U,V,W]/(V^2-UW)$. But $R(\mathbb{P}^1) = k[X,Y]$. Hence even though $\varphi(\mathbb{P}^1) \cong \mathbb{P}^1$, $R(\varphi(\mathbb{P}^1)) \ncong R(\mathbb{P}^1)$.

Conclusion. The homogeneous coordinate ring of a projective variety is not an invariant of its ismorphism class.

The situation is radically different for the affine case:

Theorem (3.9). Let Y be an affine variety. Then in the notation of (3.7), $A(Y) \cong \mathcal{O}(Y)$ so in particular $A(Y)$ is an invariant of the isomorphism class of Y.

Proof. See e.g. Hartshorne [52], page 17.

Q.E.D.

Remark (3.10). Theorem (3.9) has a very important consequence. Namely, let A be finitely generated k-algebra. Then A may be represented in the form $k[X_1,\ldots,X_n]/I$, and thus A defines an affine variety $Y = V(I)$. Moreover, from (1.11), the coordinate ring of Y will be $A(Y) = k[X_1,\ldots,X_n]/\sqrt{I}$.

Now (3.9) implies that given any other representation of A as a quotient of a polynomial algebra, if we apply the same construction, we will get the same variety (up to isomorphism).

In the projective case we have seen that the homogeneous coordinate ring is not an invariant, so that leaves us with the ring of regular functions. Unfortunately, this ring is not very interesting:

Theorem (3.11). Let Y be an irreducible projective variety. Then the only global regular functions on Y are the constant functions, i.e., $\mathcal{O}(Y) \cong k$.

Proof. For a pretty algebraic proof see [52], pages 18-19. In case $k = \mathbb{C}$, and in case Y has the structure of complex manifold we would like to indicate here an analytic proof.

The fact that Y is projective implies that Y is compact (in the complex topology; note that $\mathbb{P}^n_{\mathbb{C}}$ is compact in this topology). Now in the complex topology, it is easy to see that any regular function on Y is holomorphic. Therefore we must show that there cannot exist any global non-constant holomorphic functions on Y.

But Y being compact, given a holomorphic function $f : Y \to \mathbb{C}$, $|f|$ must have a maximum point, say at $y \in Y$. Let $U \ni y$ be a coordinate neighborhood of y (i.e. U is homeomorphic to a polycylinder). By the maximum principle for holomorphic functions, f must be constant on U. But then clearly, f is constant on all of Y.

Q.E.D.

Finally we would like to define a notion of dimension for our varieties. This is done through so-called "rational functions" :

Definition (3.12). Let Y be any irreducible variety (quasi-affine or quasi-projective). Let $U,V \subseteq Y$ be non-empty open subsets, f,g regular functions on U and V respectively. Then we say (f,U) is equivalent to (g,V) (written $(f,U) \sim (g,V)$ if $f = g$ on $U \cap V$. Note $U \cap V \neq \emptyset$ by the irreducibility of Y.

Set $K(Y) := \{$equivalence classes of pairs $(f,U)\}$.
$K(Y)$ is called the function field of Y, and the elements of $K(Y)$ are called rational functions.

Remarks (3.13).

(i) From the terminology of (3.12) we are obliged to say why $K(Y)$ is a field. It is clear that $K(Y)$ is a ring, and so we must show that every non-zero element has an inverse. Accordingly let (f,U) be a pair as in (3.12), $f \neq 0$. Then $V(f) \cap U$ is a proper closed subset of U, and $1/f$ is invertible in $V := U - V(f) \cap U$. Then the equivalence class of $(1/f,V)$ is the inverse of the equivalence class of (f,U).

(ii) The term "rational functions" comes from the fact that $K(\mathbb{A}^n) \cong K(\mathbb{P}^n) = k(X_1,\dots,X_n)$: (Proof left as exercise).

(iii) $K(Y)$ is a transcendental extension of k of finite degree.

(iv) Given Y as above, and U any Zariski open subset of Y, we have that $K(Y) \cong K(U)$.

Definition (3.14). Let Y be an irreducible variety. Then the dimension of Y is defined to be the transcendence degree of $K(Y)$ over k. Let Y be an arbitrary variety. From (1.12), (iii), Y may be written as a finite union of irreducible varieties (called the irreducible components of Y), say

$Y = Y_1 \cup \ldots \cup Y_s$ with each Y_i irreducible. Then $\dim Y := \max_i \dim Y_i$.

Remark (3.15). It follows from elementary dimension theory (Atiyah-MacDonald [6], Chapter 11) that if Y is an irreducible affine variety with coordinate ring $A(Y)$, then Krull dimension of $A(Y)$ = dimension of Y.

Exercise (3.16). Show for Y irreducible, $\dim Y = 0$ if and only if Y is a point.

Exercise (3.17). If you have failed to do Exercise (1.10) before, with our present knowledge we can easily write down a quasi-affine variety which is not affine. The standard example is $Y = \mathbb{A}^2 - \{(0,0)\}$. (Hint: First show that $\mathcal{O}(Y) \cong k[X_1,X_2]$ and then use the fact, which is clear from our above discussion, that two affine varieties are isomorphic if and only if their coordinate rings are isomorphic.)

§4. The Constructibility Theorem of Chevalley

Let X, Y be varieties, $f : X \to Y$ a morphism. In this section we would like to understand in general what is the topological structure of the image $f(X)$. We will see that this is essential for the proof of the "closed orbit lemma" (1.7) of Part III of these notes which in turn will be essential when we study the moduli of linear dynamical systems in Part IV. The problem is that $f(X)$ may be neither open or closed:

Example (4.1). Let $\mathbb{A}^3$ have coordinate ring $k[X_1,X_2,X_3]$ and let $\mathbb{A}^2$ have coordinate ring $k[X_1,X_2]$. Define $Y := V(X_1X_3-X_2) \subseteq \mathbb{A}^3$ an irreducible affine variety. Let $\pi : \mathbb{A}^3 \to \mathbb{A}^2$ be the projection map $(X_1,X_2,X_3) \longmapsto (X_1,X_2)$. Then $\pi|Y : Y \to \mathbb{A}^2$ is a morphism and the image $\pi(Y)$ is neither open nor closed. Indeed it is easy to check that $\pi(Y)$ is the union of the Zariski open subset $\mathbb{A}^2 - V(X_1)$ and the origin $(0,0)$.

This leads us to make the following definitions:

Definition (4.2).

(i) Let X be any topological space. Then a subset $Z \subset X$ is locally closed if it is the intersection of an open subset and a closed subset of X.

(ii) X as in (i). Then a subset $Z \subset X$ is constructible if Z may be expressed as a disjoint finite union of locally closed subsets of X.

We state the following theorem without proof. Complete proofs may be found in Humphreys [71], page 33, and in Altman-Kleiman [2], pages 97-99.

Theorem (4.3.). (Chevalley) *Let $f : X \to Y$ be a morphism of varieties. Let $Z \subseteq X$ be a constructible subset. Then $f(Z)$ is constructible.*

Remark (4.4). In Section 3 above we mentioned the notion of completeness and said that $\mathbb{P}^n$ is complete. One of the characteristic properties of complete varieties is that images of closed subsets are closed. For details about this see Mumford [106] and [107].

§5. Sheaves and Vector Bundles

We give here a very brief sketch of the facts about vector bundles and sheaves which we will need in what follows. The interested reader is referred to Serre [125] or Hartshorne [52] for further details.

Sheaves are used in mathematics to relate local data to global data. The exact definition is:

Definitions (5.1).

(i) Let X be a topological space. A presheaf of abelian groups consists of (a) a correspondence which to every open set U associates an abelian group $\mathcal{F}(U)$ with $\mathcal{F}(\emptyset) = 0$; (b) for every inclusion $V \subseteq U$ of open sets there exists a restriction homomorphism of abelian groups ρ_{UV}: $\mathcal{F}(U) \to \mathcal{F}(V)$ such that ρ_{UU} is the identity, and for $W \subseteq V \subseteq U$, $\rho_{UW} = \rho_{VW} \circ \rho_{UV}$.

(ii) A sheaf of abelian groups $\mathcal{F}$ on X is a presheaf which obeys the following two axioms:

(a) Identity: Given s, $s' \in \mathcal{F}(U)$, and any open cover $\{U_i\}$ of U such that $\rho_{UU_i} s = \rho_{UU_i} s'$ for all U_i, then $s = s'$.

(b) Gluability: Given an open U, and any open covering $\{U_i\}$ of U suppose that $s_i \in \mathcal{F}(U_i)$ are such that $\rho_{U_iU_i \cap U_j} s_i = \rho_{U_jU_j \cap U_i} s_j$ for all i and j. Then there exists $s \in \mathcal{F}(U)$ such that $\rho_{UU_i} s = s_i$.

Remarks-Definitions (5.2).

(i) Clearly using the same definitions (5.1) above, we can define notions of presheaves and sheaves of sets, rings, etc.

(ii) Elements of $\mathcal{F}(U)$ are usually called sections. Given a section $s \in \mathcal{F}(U)$ and an open subset $V \subseteq u$, the restriction $\rho_{UV} s$ is usually written as $s|V$.

(iii) For $\mathcal{F}$ a presheaf on X, the stalk of $\mathcal{F}$ at $p \in X$, denoted by $\mathcal{F}_p$, is $\varinjlim_{U \in p} \mathcal{F}(U)$ (the direct limit [6] over all open $U \ni p$ defined by the restriction maps).

Example (5.3). Let X be a variety. For each open $U \subset X$, let $\mathcal{O}(U)$ be the ring of regular functions on U. Then the correspondence $U \to \mathcal{O}(U)$ defines a sheaf on X, denoted by $\mathcal{O}_X$. $\mathcal{O}_X$ is called the structure sheaf of X.

Now we will see that sheaves generalize the classical notion of vector bundle. To see this we must first make the following definitions:

Definitions (5.4).

(i) Let X be a variety and $\mathcal{O}_X$ its structure sheaf. An $\mathcal{O}_X$-module $\mathcal{F}$ is a sheaf on X such that for each open set $U \subseteq X$, $\mathcal{F}(U)$ is an $\mathcal{O}_X(U)$-module, and for each open $V \subseteq U$, the restriction $\mathcal{F}(U) \to \mathcal{F}(V)$ is compatible with the module structures via the ring homomorphism $\mathcal{O}_X(U) \to \mathcal{O}_X(V)$.

(ii) Given a sheaf $\mathcal{F}$ on X we can clearly restrict the sheaf to any open subset U and get a sheaf $\mathcal{F}|U$ on U. (See e.g. [125].) Then an $\mathcal{O}_X$-module $\mathcal{F}$ is said to be locally free if there exists an open covering of X by open sets U such that $\mathcal{F}|U$ is a free $\mathcal{O}_X|U$-module. The definition of the rank of $\mathcal{F}$ over U should be clear.

(iii) An algebraic vector bundle of rank n over X is a variety $\mathcal{B}$, and a morphism $\pi: \mathcal{B} \to X$, such that there exists an open covering $\{U_i\}$ of X and isomorphisms $\varphi_i: \pi^{-1}(U_i) \to k^n \times U_i$ (called trivializations) with the property that for any i,j and for any open affine $V \subset U_i \cap U_j$ the automorphism $\varphi_{ij} = \varphi_i \circ \varphi_j^{-1}$ of $k^n \times V$ is linear, i.e. if we identify for each $x \in X$, $\pi^{-1}(x)$ with k^n, then $\varphi_{ij}(x) \in GL(n,k)$ and this correspondence is a morphism. (Here $GL(n,k)$ denotes the general linear group. See Part III, (1.2)(ii).)

Examples (5.5).

(i) Let A be a finitely generated k-algebra without nilpotent elements (i.e. A is reduced). Then as we have seen in (3.10) one can associate a variety X unique up to isomorphism to A. Let M be an A-module. Then to M we can associate a sheaf of $\mathcal{O}_X$-modules ($\mathcal{O}_X$ = structure sheaf of X) by setting $\tilde{M} := \mathcal{O}_X \otimes_A M$.

A word of explanation is in order here. We consider A as a topological ring with A having the discrete topology. Then A defines a constant sheaf of rings (also denoted by A) on X by setting for each $U \subset X$ open, $A(U) :=$ ring of continuous functions from $U \to A$. Similarly M defines a constant sheaf of modules over X, and in the obvious way M will be a sheaf of A-modules.

Now given any variety V, if $\mathcal{F}$ and $\mathcal{G}$ are sheaves of $\mathcal{O}_V$-modules one can define a presheaf on V by the rule $U \mapsto \mathcal{F}(U) \otimes_{\mathcal{O}_V(U)} \mathcal{G}(U)$ for $U \subset V$ open. This in general will not be a sheaf. However, given any presheaf $\mathcal{H}$ on any topological space Y, there is a canonical procedure for constructing an associated sheaf $\mathcal{H}^a$, unique up to isomorphism, with the property that $\mathcal{H}^a_p \cong \mathcal{H}_p$ for all $p \in V$. (For this construction see e.g.

Serre [125].) Then $\mathcal{F} \otimes_{\mathcal{O}_V} \mathcal{G}$ denotes the sheaf associated to the presheaf $U \mapsto \mathcal{F}(U) \otimes_{\mathcal{O}_V(U)} \mathcal{G}(U)$. Finally $\tilde{M} = \mathcal{O}_X \otimes_A M$ is taken in this sheaf-theoretic sense.

Let $x \in X$ be a point. Then x corresponds to a maximal ideal in A which we will also denote by x. We leave it as an exercise for the reader to check that $\tilde{M}_x \cong M_x := M \otimes_A A_x$ ($A_x := A$ localized at x). In particular if M is a finitely generated projective A-module, M_x is free for all $x \in X$ (see Serre [127], and hence $\tilde{M}$ is a locally free $\mathcal{O}_X$-module.

(ii) Let $\pi: \mathcal{V} \to X$ be a vector bundle over the variety X. The fiber of $\mathcal{V}$ over $x \in X$ is just $\mathcal{V}_x := \pi^{-1}(x)$. Then associated to the vector bundle $\mathcal{V}$ we have the dual bundle $\pi^* : \mathcal{V}^* \to X$ such that $\mathcal{V}_x^* := \mathrm{Hom}_L(\mathcal{V}_x, k)$. It is easy to show that the $\mathcal{V}_x^*$ fit together to form a vector bundle $\mathcal{V}^*$.

(iii) We will use some terminology here that we have not defined such as "smooth" and "tangent space". We believe it is intuitively clear what we mean by these terms and for the formal definitions the reader is referred to Mumford [107], pages 3-9. We remark that over $\mathbb{C}$, to say that a variety is smooth, means that it has the structure of a complex manifold. Then the algebraic tangent space (the so-called "Zariski tangent space") is exactly the usual complex tangent space from differential geometry (Griffiths-Harris [45]).

Let X be a smooth irreducible variety of dimension n. Let $T_X := \bigcup_{x \in X} T_{X,x}$ where $T_{X,x}$ = tangent space of X at x. Then T_X may be given the structure of algebraic variety. Let $\pi : T_X \to X$ be defined by setting $\pi(T_{X,x}) = x$. Then this gives T_X the structure of a vector bundle of rank n over X, called the tangent bundle. If X is a curve (i.e. has dimension 1), then T_X has rank 1. In general a rank 1 vector bundle over a variety is called a line bundle.

Remark (5.6). Notice from (5.4)(iii) that for a sufficiently fine affine open covering $\mathfrak{U} = \{U_i\}_{i \in I}$ of X, the vector bundle $\mathcal{V}$ is determined by the morphisms $\varphi_{ij}: U_i \cap U_j \to GL(n,k)$ defined by $\varphi_{ij}(x) := (\varphi_i \circ \varphi_j^{-1})|_{\{x\} \times k^n}$ (notation as in (5.4)(iii)), and these morphisms satisfy the cocycle condition that $\varphi_{ij} \circ \varphi_{jk} = \varphi_{ik}$ on $U_i \cap U_j \cap U_k$ for all $i,j,k \in I$. One also says that $\{\varphi_{ij}\}$ defines a Čech cocycle of the sheaf of $\mathcal{O}_X$-module automorphisms of $\mathcal{O}_X^n$, denoted by $GL(n, \mathcal{O}_X)$, with respect to the covering $\mathfrak{U}$. This Čech cocycle clearly determines the vector bundle $\mathcal{V}$.

Now one can clearly take refinements of the covering $\mathfrak{U}$, and find appropriate Čech cocycles which also define $\mathcal{V}$. Moreover there is an equivalence relation among Čech cocycles defined by a coboundary condition which identifies isomorphic vector bundles. In technical language, vector bundles of rank n over X are classified by the Čech cohomology pointed set $\check{H}^1(X, GL(n, \mathcal{O}_X))$. We shall not go into details here about this. Good references about all this are Gunning [47], [48], and Serre [125], [128] (this last reference for details about non-abelian cohomology).

The connection between sheaves and vector bundles is given by the following fundamental theorem:

Theorem (5.7). Let X be a variety. Then there exists a canonical 1-1 correspondence between isomorphism classes of locally free sheaves on X of rank n and isomorphism classes of rank n vector bundles.

Proof. We sketch the proof here. For details see [52] or [106].

We will show how given a vector bundle $\pi: \mathcal{V} \to X$, one may derive a locally free sheaf. Let $U \subseteq X$ be any open subset. Recall that a section over U is a morphism $\sigma: U \to \mathcal{V}$ such that $\pi \circ \sigma =$ identity on U. Then one can easily show the correspondence $U \longrightarrow \{\text{sections over } U\}$ is a sheaf, $\mathcal{J}(\mathcal{V})$. One must check then that the vector bundle structure $\pi: \mathcal{V} \to X$ gives that $\mathcal{J}(\mathcal{V})$ is locally free ([152], page 129).

Conversely, given a locally free sheaf $\mathcal{S}$ on X, using the symmetric algebra defined by $\mathcal{S}$ one can define a vector bundle on X, $\mathcal{V}_{\mathcal{S}}$ ([152], page 128).

Finally one can show that $\mathcal{J}(\mathcal{V}_{\mathcal{S}}) \cong \underline{Hom}(\mathcal{S}, \mathcal{O}_X)$ (the dual sheaf of $\mathcal{S}$; see [52], pages 123 and 129). This establishes the 1-1 correspondence.

Q.E.D.

Remarks (5.8).

(i) Suppose that X is an affine variety with coordinate ring R. Let $\mathcal{V}$ be a vector bundle over X. Then to $\mathcal{V}$ we can associate an R-module M, by setting $M := \Gamma(X, \mathcal{V}) := \{\text{global sections of } \mathcal{V} \text{ over } X\}$. The module structure on M is the obvious one. Now M is a locally free R-module by (5.7) i.e., M is projective. Notice that under the correspondence of (5.7) $\mathcal{V}^*$ corresponds to $\widetilde{M}$ (notation as in (5.5)(i)) and we shall usually identify these objects in the sequel.

(ii) Perhaps the most famous conjecture of Serre is that any finitely generated projective R-module is free when $R = k[X_1, \ldots, X_n]$ is a polynomial ring. This conjecture has had a long history with many attempts to prove it and several partial results (e.g., the case $n = 2$ is due to Seshadri [129]). Finally in 1975 almost simultaneously complete proofs were given independently by D. Quillen [116] and A. Suslin [144]. Consequently in the sequel this important result will be referred to as the Quillen-Suslin theorem. Note that by (5.7) this means that any vector bundle of rank m over $\mathbb{A}^n$ is trivial i.e. isomorphic to $\mathbb{A}^n \times k^m$.

Finally, in Part VIII we shall need the following definition (Serre [125]):

Definition (5.9). Let $\mathbb{P}^n$ be a projective space with homogeneous coordinate ring $k[X_0,\ldots,X_n]$. As in (2.8), let $U_i = \{(a_0,\ldots,a_n) \in \mathbb{P}^n \mid a_i \neq 0\}$ and let $\mathcal{O} := \mathcal{O}_{\mathbb{P}^n}$ be the structure sheaf on $\mathbb{P}^n$. Set $\mathcal{O}_i := \mathcal{O}|U_i$ for each $i = 0,\ldots,n$. On $U_i \cap U_j$, the function X_j/X_i is regular with values in k^*. Hence for any $m \in \mathbb{Z}$, multiplication by the function X_j^m/X_i^m defines an isomorphism $\lambda_{ij}(m): \mathcal{O}_j|U_i \cap U_j \to \mathcal{O}_i|U_j \cap U_i$. Clearly the $\lambda_{ij}(m)$ have the property that $\lambda_{ii}(n) = \text{identity}$, and $\lambda_{ik}(m) = \lambda_{ij}(m) \circ \lambda_{jk}(m)$ (the cocycle condition) on $U_i \cap U_j \cap U_k$. We leave it as an easy exercise (or see e.g. Serre [125]) to show that this implies that there exists a unique sheaf on $\mathbb{P}^n$, denoted by $\mathcal{O}(n)$, together with isomorphisms $\alpha_i: \mathcal{O}(m) \mid U_i \xrightarrow{\sim} \mathcal{O}_i$ such that for every i,j, $\alpha_j = \lambda_{ij}(m) \circ \alpha_j$ on $U_i \cap U_j$. $\mathcal{O}(n)$ is said to be formed by glueing the $\mathcal{O}_i$ together via the $\lambda_{ij}(m)$.

Remark (5.10). It is a fundamental theorem of Birkhoff-Grothendieck [8], [46] that any vector bundle of rank n over $\mathbb{P}^1$ is isomorphic to a direct sum of line bundles $\bigoplus_{i=1}^{m} \mathcal{O}_{\mathbb{P}^1}(n_i)$. Since we will interpret this result in a system theoretic context we will defer its proof until Part VIII, Section 2.

§6. The Grassmann Variety

The construction of Kalman of a quotient space of completely reachable matrix pairs under the natural action of the general linear group uses a technique that is a straightforward generalization of the construction of the Grassman variety. (We will discuss this in Part IV, Section 2. For Kalman's original construction see [79].) Therefore it seems worthwhile to include here a brief look at the Grassmann variety. For a beautiful modern treatment of Grassmannians see Griffiths-Harris [45], pages 193-211. Another nice treatment is in Chern [24], pages 64-79.

As is probably well-known to most of the readers we define for $m \leq n$

$$Gr(m,n) \overset{\text{point set}}{=} \{m\text{-planes passing through the origin in } k^n\}.$$

Of course as point sets $Gr(1,n) = \mathbb{P}^{n-1}$, and as for $\mathbb{P}^{n-1}$ we want to give the Grassmannian $Gr(m,n)$ the structure of algebraic variety.

Let $\Lambda \in Gr(m,n)$. Then we may find m linearly independent vectors $v_1,\ldots,v_m \in k^n$ which span Λ. Writing each vector as an $1 \times n$ row vector relative to the standard basis of k^n we can express Λ in matrix form as

$$A = \begin{bmatrix} v_{11} & \cdots & v_{1n} \\ \vdots & & \\ v_{m1} & \cdots & v_{mn} \end{bmatrix}$$

where A has rank m. Given any two matrices A, A' of rank m, A and A'

represent the same m-plane Λ if and only if there exists $g \in GL(m,k)$ such that $A = gA'$. Thus if we denote by $M^{reg}_{m,n}$ the space of $m \times n$ matrices of maximal rank we have that $Gr(m,n) \cong M^{reg}_{m,n}/GL(m,k)$ as point sets. (Compare the construction in Part IV, Section 2.)

Let I be a multi-index contained in the set $\{1,\dots,n\}$ of size m. Let

$U_I = \{\Lambda \in Gr(m,n) \mid \Lambda$ admits a matrix representation such that the I-th $m \times m$ minor of A is non-singular$\}$.

Then any $\Lambda \in U_I$ admits a unique matrix representation such that the I-th $m \times m$ minor is the identity. Conversely any matrix A whose I-th $m \times m$ minor is the identity will represent an element of U_I. Thus the set U_I depends on $m(n-m)$ free parameters i.e. there exists a bijection

$$\pi_I: U_I \to k^{m(n-m)} .$$

One can check that for any multiindices I and I', $\pi_I(U_I \cap U_{I'})$ is open in $k^{m(n-m)}$ and $\pi_{I'} \circ \pi_I^{-1}$ is a morphism on this open subset. This means that the U_I patch together to form the smooth algebraic variety $Gr(m,n)$. Now $Gr(m,n)$ is even a projective variety. This is the classical Plücker embedding. Indeed define the <u>Plücker map</u>

$$\varphi: Gr(m,n) \to \mathbb{P}(\Lambda^m k^n) \cong \mathbb{P}^{\binom{n}{m}-1}$$

as follows: If $\Lambda \in Gr(m,n)$ is spanned by $v_1,\dots,v_m$, then send $\Lambda \longmapsto v_1 \wedge \dots \wedge v_m$. In terms of the standard basis $e_1,\dots,e_n$ of k^n we may write a basis of $\Lambda^m k^n$ as $\{e_{i_1} \wedge \dots \wedge e_{i_m}\}_I$ over all multiindices $I = \{i_1 < \dots < i_m\} \subset \{1,\dots,n\}$ and then the Plücker map is defined by sending $\Lambda \longmapsto (\dots,\det \Lambda_I,\dots)$ where Λ_I is the I-th $m \times m$ minor of a matrix representation of Λ. It is easy to check then that φ in point of fact is an embedding. (See [45], page 209.)

What is not so trivial is to prove that $\varphi(Gr(m,n)) \hookrightarrow \mathbb{P}^{\binom{n}{m}-1}$ is cut out by quadrics (the "Plücker relations"). For a very pretty proof of this see [45], pages 210-211.

We conclude this section with a definition and some remarks which we will need in Part IV, Section 5.

<u>Definition (6.1).</u> Let $\pi: E \to B$ be a continuous mapping of topological spaces, and G a topological group. Then (E,p) is called a <u>topological principal G-bundle</u> if there exists an open covering $\mathfrak{U} = \{U_i\}$ of B and homeomorphisms $\varphi_i: \pi^{-1}(U_i) \to U_i \times G$ which map $\pi^{-1}(x)$ into $x \times G$ for all $x \in U_i$ such that the maps $\varphi_{ij}: U_i \cap U_j \to G$ given by $\varphi_{ij}(x) = (\varphi_i \circ \varphi_j^{-1})|_{\{x\}\times G}$ are continuous (where we identify $\{x\} \times G$ and G). We leave it as an exercise for the reader to define notions of <u>algebraic principal bundle</u> (for the definition of <u>algebraic group</u>, see

Part III, (1.1)) and over $\mathbb{C}$ holomorphic principal bundle.

Now consider once again the morphism $\pi: M_{m,n}^{reg} \to Gr(m,n)$ exhibiting $Gr(m,n)$ as the space of orbits of $M_{m,n}^{reg}$ under the action of $GL(m,k)$. Clearly the fibers are isomorphic to $GL(m,k)$ and it is an easy exercise (which we will carry out explicitly in Part IV, (5.1) (ii)) to show that $(M_{m,n}^{reg},\pi)$ is an algebraic principal $GL(m,k)$-bundle. This simple observation will be crucial in our discussion of the moduli of linear dynamical systems.

§7. Some Remarks on Algebraic Geometry Over a Non-Algebraically Closed Field

Many times in system theory one finds that the natural field of definition is not algebraically closed e.g. in Part VIII we shall be interested in the field of real numbers $\mathbb{R}$. Finite fields have also been useful in system theory via automata theory [32]. Hence even though for us the case of k being algebraically closed is the most important case, we should like to conclude our sketch of algebraic geometry with some remarks about varieties defined over non-algebraically closed fields.

Throughout the discussion we will take k to be an algebraically closed field complete, and k' an arbitrary subfield. For complete details about this topic see Humphreys [71] and Mumford [106]. See also our discussion in Part VI, Section 6.

Fields of Definition (7.1).

(i) Let $X \subset \mathbb{A}_k^n$ be an affine variety with ideal $I(X)$. If $I(X) = (f_1,\ldots,f_r)$, $f_i \in k'[X_1,\ldots,X_n]$ we will say that X is defined over k' .

(ii) Let $R' := k'[X_1,\ldots,X_n]$ and $R := R' \otimes_{k'} k \cong k[X_1,\ldots,X_n]$. If $R'_X := R'/I(X)$, then the coordinate ring of X , $R_X \cong R'_X \otimes_{k'} k$. Since the coordinate ring of X is intrinsically defined, this gives an intrinsic characterization to the notion that X is defined over k' . Using the fact that any variety is covered by affine open subsets, it is easy to generalize the definition of "defined over k' " to arbitrary k-varieties. (The definition is left as an exercise or see [71] and [106].) We however shall stick to the affine case here.

(iii) If now $X \subset \mathbb{A}^n$ is any variety, we let $X(k') = X \cap k'^n$ and call this the set of k'-rational points. Given a morphism $\varphi: X \to Y$ of varieties defined over k' , φ is said to be defined over k' if its local coordinate functions all have coefficients in k' .

Now the extra structure that X derives by being defined over k' can be explicitly expressed as follows:

k'-Topology (7.2).

(i) Let $Gal(k/k')$ be the Galois group of automorphisms of k which fix k' . Then if X is as in (7.1)(i) for each $\lambda \in Gal(k/k')$, one has a k-algebra homomorphism $\lambda^{-1}: R_X \to R_X$ defined by $\lambda^{-1}(f + I(X)) := f^{\lambda^{-1}} + I(X)$ where $f^{\lambda^{-1}}$ denotes the element of R (notation as in (7.1)) gotten by applying λ^{-1} to the coefficients of $f \in R$. It is easy to check that if $\lambda_X: X \to X$ is the corresponding morphism of varieties, then $\lambda_X((a_1,\ldots,a_n)) = (\lambda(a_1),\ldots,\lambda(a_n))$ for $(a_1,\ldots,a_n) \in X \subset \mathbb{A}_k^n$. Moreover for $\lambda,\mu \in \mathrm{Gal}(k/k')$ we have $(\lambda \circ \mu)_X = \lambda_X \circ \mu_X$ so that this construction defines a natural action of $\mathrm{Gal}(k/k')$ on X.

(ii) The k'-<u>topology</u> on X is the set of open $\mathrm{Gal}(k/k')$ invariant subsets.

(iii) We now consider $\mathbb{A}_k^n$ with the k'-topology. Then to say a variety $Y \subset \mathbb{A}_k^n$ is k'-closed, is weaker than to say that Y is defined over k'. Indeed we have the following proposition which we state without proof (see Mumford [106], page 194):

<u>Proposition (7.3).</u> <u>If</u> $Y \subset \mathbb{A}_k^n$ <u>is</u> k'-<u>closed, then</u> Y <u>is defined over a finite, purely inseparable extension of</u> k'. <u>In particular if</u> k' <u>is perfect, then</u> Y <u>is</u> k'-<u>closed if and only if it is defined over</u> k'.

For some nice examples of the previous theory on [106], pages 194-197.

PART II. SOME BASIC SYSTEM THEORY

This part is intended to introduce the mathematician with little or no background in system and control theory to the basic definitions and ideas of the subject. No attempt at completeness will be made.

In the literature now there are several good treatments of the topics we intend to cover here. Perhaps the best for the subject matter of these notes are the works of Kalman-Arbib-Falb [82], and Kalman [75] which consider system theory from a modern algebraic point of view. There is also the recent book of Luenberger (93] which contains many illustrative examples. For a more classical treatment there is the very good book of Jacobs [74]. Of course all these references contain more exhaustive bibliographies for the interested reader.

One feature of our treatment here which is not usually considered in the standard references is the inclusion of a beautiful argument of Deligne [26] concerning the equivalence of the differential equation and state space formulation of certain kinds of dynamical systems. This section (Section 4) will include some elementary differential geometry (the techniques of which are also useful in system theory) and except for the results in the time-invariant case, may be skipped by the reader.

§1. Dynamical Systems

In this section we give the basic elements from the theory of dynamical systems which will be used throughout these notes. Our treatment is based on that of Kalman-Arbib-Falb [82].

The essential point is to make precise the notion of a "device" Σ which accepts inputs $u(t)$ and transforms them into outputs $y(t)$. This transformation may take place at discrete intervals (e.g. a digital computer) or continuously over time. The "state" of Σ is then that part of the present and past history of Σ which is needed to determine the outputs.

The exact definition is:

Definition (1.1). A dynamical system Σ is specified by the following axioms:

(i) A time set T which is an ordered subset of $\mathbb{R}$ (the real numbers), a state set X, a set of input values U, a nonempty set of input functions $\Omega = \{u: T \to U\}$, a set of output values Y, and a set of output functions $\Gamma = \{\gamma: T \to Y\}$.

(ii) For $u \in \Omega$, set $u_{(t_1,t_2]} := u \,|\, (t_1,t_2] \cap T$. Then for $u,u' \in \Omega$, if

$t_1 < t_2 < t_3$, there exists $u'' \in \Omega$ such that $u''_{(t_1,t_2]} = u_{(t_1,t_2]}$ and $u''_{(t_2,t_3]} = u'_{(t_2,t_3]}$. This is called concatenation of inputs.

(iii) There is a state transition function $\varphi: T \times T \times X \times \Omega \to X$ whose value $x(t) := \varphi(t;\tau,x,u)$ is the state resulting at time t from the initial state $x(\tau)$ under the action of the input u. φ has the following properties:

(a) φ is defined for $t \geq \tau$, but not necessarily for $t < \tau$.

(b) $\varphi(t;t,x,u) = x$ for any $t \in T$, $x \in X$, $u \in \Omega$.

(c) For $t_1 < t_2 < t_3$, $\varphi(t_3;t_1,x,u) = \varphi(t_3;t_2,\varphi(t_2;t_1,x,u),u)$ for all $x \in X$, $u \in \Omega$. (This is sometimes called the semigroup property.)

(d) For $u,u' \in \Omega$ with $u_{(\tau,t]} = u'_{(\tau,t]}$, $\varphi(t;\tau,x,u) = \varphi(t;\tau,x,u')$.

(iv) There is a readout map $\eta: T \times X \to Y$ which gives the output $y(t) = \eta(t,x(t))$ at time $t \in T$.

Remarks (1.2).

(i) In the literature the set $T \times X$ is usually called the event space and elements of $T \times X$ are called events. The state transition function is referred to as trajectory, motion, flow, etc.

(ii) Definition (1.1) is considered to be an internal description of a system Σ since it is characterized by describing the systems through its states which are clearly internal and rather abstract attributes. Many times however a system is characterized through its input/output behavior. In the case of a system Σ as described in (1.1), we can get an input/output function by defining for an event $(\tau,x) \in T \times X$,

$$f_{\tau,x}(u_{(\tau,t_1]})(t) := \eta(t,\varphi(t;\tau,x,u))$$

where $t \in (\tau,t_1]$.

We want to have therefore a definition of a "dynamical system" from an input/output or external point of view:

Definition (1.3). A dynamical system Σ in the input/output (or external) sense is specified by the following axioms:

(i) There are sets T, U, Ω, Y, Γ satisfying all the properties of (1.1).

(ii) There is a set I indexing a family of functions $\mathfrak{F} := \{f_i : T \times \Omega \to Y, i \in I\}$ such that for each $f_i \in \mathfrak{F}$ (an input/output function), there is a map $\lambda: I \to T$ such that $f_i(t,u)$ is defined for $t \geq \lambda(i)$ and if $\tau < t$ where $\tau,t \in T$, and $u,u' \in \Omega$ with $u_{(\tau,t]} = u'_{(\tau,t]}$, then $f_i(t,u) = f_i(t,u')$ for all i such that $\tau = \lambda(i)$.

Remarks (1.4).

(i) The realization problem consists of constructing a dynamical system in the sense of (1.1) from one in the sense of (1.3). In other words we want to construct an internal description (1.1) from the external input/output behavior.

Actually, abstractly, this is trivial and is a well-known construction in automata theory which we sketch here. When our systems have more algebraic and analytic structure, the problem becomes much less trivial since we will demand realizations of input/output maps satisfying certain specified conditions, and this will be the topic of Part VI of these notes.

What we do here is to construct a realization of (1.3) on the set theoretic level. We will construct the state space X and leave it as a simple exercise for the reader to construct φ and η (notation as in (1.1)). First looking at the discussion in (1.2),(ii), it is clear that the index set I must correspond to a subset of $T \times X$. Now I may be too small and so we first enlarge $\mathfrak{F}$ by including all functions g defined by $g(t,u) = f_i(t,u')$ where $\tau < t_1 < t_2$, $u_{(t_1,t_2]} = u'_{(t,t_2]}$, $t \in (t_1,t_2]$, and $\tau = \lambda(i)$. Next set $g = f_j$ $(j \notin I)$ and put $\lambda(j) = t_1$, and replace I by $I \cup \{j\}$. The state space is then $X = \bigcup_{\tau \in T} X_\tau$ where $X_\tau = \{i \in I \mid \lambda(i) = \tau\}$. (For details about this see [82] especially Chapter 7 which also contains a set of references for the relevant automata theory literature.)

(ii) In order to give some meat to our discussion we must make our definitions somewhat more concrete which will be our project now.

<u>Definitions (1.5).</u>

(i) A dynamical system Σ (in the sense of (1.1)) is <u>continuous</u> if $T = \mathbb{R}$, and <u>discrete</u> if $T = \mathbb{Z}$.

(ii) Σ is <u>smooth</u> if:

(a) Σ is continuous.

(b) X, Ω are topological vector spaces.

(c) The map $(\tau,x,u) \longmapsto \varphi(\cdot;\tau,x,u)$ defines a continuously differentiable map from $T \times X \times \Omega \to$ {continuously differentiable functions from $T \to X$}.

<u>Proposition (1.6).</u> <u>For Σ smooth, φ is a solution of a differential equation</u>

$$\frac{dx}{dt} = f(t,x,u) .$$

<u>Proof.</u> Immediate from (1.5)(ii)(c).

Q.E.D.

<u>Definitions (1.7).</u>

(i) Σ (as in (1.1)) is <u>linear</u> if:

(a) X, Ω, Y, Γ are vector spaces over an arbitrary field k.

(b) $\varphi(t;\tau,\cdot,\cdot) : X \times \Omega \to X$ is k-linear for all t and τ.

(c) $\eta(t,\cdot) : X \to Y$ is k-linear for all t.

(ii) Σ is <u>finite dimensional</u> if X is a finite dimensional vector space, in which case <u>dimension</u> of Σ : = dimension of X .

<u>Remark (1.8).</u> It is an easy exercise (see [82], pages 30-31) to verify that given a smooth linear finite dimensional system Σ of dimension n , such that the space of input values $U \cong k^m$ and the space of output values $Y \cong k^p$ with $k = \mathbb{R}$ or $\mathbb{C}$, then Σ may be represented as

$$\dot{x}(t) = F(t)x(t) + G(t)u(t)$$

$$y(t) = H(t)x(t)$$

where $x^t(t) := (x_1(t),\ldots,x_n(t))$ is a state vector C i.e. $x(t)$ is a column vector and $x^t(t)$ is the transpose), $\dot{x}^t(t) := (\frac{dx_1}{dt},\ldots,\frac{dx_n}{dt})$, $F(t)$ is an $n \times n$ matrix of functions t , $G(t)$ $n \times m$, $H(t)$ $p \times n$, and $u(t) \in k^m$, $y(t) \in k^p$ (the input and output values, respectively, written as column vectors).

Finally we have:

<u>Definition (1.9).</u> A dynamical system Σ in the sense of (1.1) is <u>time-invariant</u> or <u>constant</u> if

(i) T is an additive group.

(ii) Ω is closed under the <u>shift operator</u> σ_τ , defined by $\sigma_\tau u(t) := u(t+\tau)$ for all $t,\tau \in T$.

(iii) $\varphi(t;\tau,x,u) = \varphi(t+s;\tau+s,x,\sigma_s u)$ for all $s \in T$.

(iv) $\eta(t,\cdot) : X \to Y$ is independent of t .

<u>Remarks (1.10).</u>

(i) From (1.8) it is immediate that if Σ is a smooth finite dimensional time-invariant system of dimension n , then Σ may be represented as

$$\dot{x} = Fx+Gu$$

$$y = Hx$$

where now the matrices F, G, H are constant.

(ii) One can play a similar game in the discrete case $(T = \mathbb{Z})$ and we leave it to the reader to verify that if Σ is a discrete linear constant finite dimensional system of dimension n , such that the space of input values $U \cong k^m$ and the space of output values $Y \cong k^p$ with k an arbitrary field, then Σ may be represented by difference equations

$$x(t+1) = Fx(t) + Gu(t)$$

$$y(t) = Hx(t)$$

where F is $n \times n$, G $n \times m$, H $p \times n$ are all constant and defined over k .

Note moreover that these difference equations make sense even for F, G, H defined over an arbitrary commutative ring with identity R , i.e., one has the notion of a <u>system over a ring</u>. The physical motivation for this will be discussed in Parts VI and VIII. Systems with k taken to be a finite field come up e.g. in the theory of linear sequential circuits.

(iii) Finally note that in the time-invariant case one may identify systems with triples of matrices (F,G,H) of appropriate sizes defined over a field k or more generally over a ring R.

§2. Controllability and Reachability

In system and control theory a basic objective is to be able to transfer the motion from one point in the state space to another via proper application of controls and to describe which points in the state space can be reached from a given point in this way. This leads to two fundamental concepts which are essential in modern system theory and which have a large literature devoted to them, namely "controllability" and "reachability".

Throughout this section we will assume for simplicity that our systems are linear even though notions of "controllability" and "reachability" can be defined in the non-linear case. In Part VI, Section 6 we will in point of fact consider such definitions for polynomial systems, but for the present we want to keep the exposition as simple as possible.

Throughout this section unless explicitly stated otherwise, Σ will denote a finite dimensional, smooth, linear dynamical system. We will use the same notation as in (1.1).

Definitions (2.1).

(i) An event (τ,x) is controllable (to the zero state) if there exists $t \geq \tau$ and a $u \in \Omega$ (t and u may of course depend on (τ,x).) such that $\varphi(t;\tau,x,u) = 0$. Σ is completely controllable if it is controllable for every event (τ,x).

(ii) An event (τ,x) is reachable (from the zero state) if there is an $s \leq \tau$ and a $u \in \Omega$ (s and u may depend on (τ,x)) such that $\varphi(\tau;s,0,u) = x$. Σ is completely reachable if it is reachable for every event (τ,x).

Remarks (2.2).

(i) In (2.3) we will see that for Σ constant, the notions of controllability and reachability coincide. In general however, (see Casti [22]), this is false. A standard counterexample is the system $dx/dt = f(t)u(t)$ where $f(t)$ is continuous and identically zero for $t < 0$. Then clearly $(0,x)$ is controllable but not reachable.

(ii) It is always true that the set of controllable states contains the set of reachable states (see [82], page 41). As remarked above in the constant case we have the following fundamental result:

Theorem (2.3). *Let* Σ *be a finite dimensional smooth linear constant dynamical system of dimension* n . *Then the following properties are equivalent:*

(i) Σ *is completely reachable.*

(ii) Σ *is completely controllable.*

(iii) *If* $\Sigma = (F,G,H)$ *(notation as in* (1.10)(iii)), *then* $\text{rank}(G\ FG\ \dots\ F^{n-1}G) = n$ *where* $(G\ FG\ \dots\ F^{n-1}G)$ *is the* $n \times mn$ *matrix made up of the columns of the matrices* $G\ FG\ \dots\ F^{n-1}G$.

Proof. The idea is to show that both properties (i) and (ii) are equivalent to (iii). We will sketch the proof that (ii) is equivalent to (iii) here and refer the reader to [82], pages 33-40 for a complete proof of this theorem.

From our remarks in (1.10) and from the elementary theory of linear constant differential equations, the state transition function φ in this constant case is of the form

$$\varphi(t;\tau,x,u) = e^{(t-\tau)F}x + \int_{\tau}^{t} e^{(t-s)F}G\,u(s)ds \ .$$

An event (τ,x) is controllable if and only if $\varphi(t;\tau,x,u) = 0$ which means that

$$(*) \quad 0 = x + \int_{\tau}^{t} e^{(\tau-s)F}\ G\,u(s)ds \ .$$

Hence we must show that for every (τ,x) there exists t and u which satisfy (*) if and only if $C: = (G\ FC\ \dots\ F^{n-1}G)$ is of maximal rank. Suppose first that rank $C < n$. Then there exists a vector $v \in X$ (the state space) such that $(v^tC) = 0$ (where v^t = transpose of v) and hence by the Cayley-Hamilton theorem $v^tF^qG = 0$ for all $q \geq 0$. We claim then that any event of the form $(0,x)$ is uncontrollable when $v^tx \neq 0$. Indeed setting $\tau = 0$ in (*) we have

$$(**) \qquad 0 = x + \int_{0}^{t} e^{-sF}\ G\,u(s)ds \ .$$

Then looking at the power series expansion of e^{-sF} we see that $v^t(\int_0^t e^{-sF}\ G\,u(s)ds) = 0$, and thus for x with $v^tx \neq 0$, $(0,x)$ can never be controllable.

Conversely suppose that rank $C = n$. Since the system is time-invariant we may always assume that $\tau = 0$ i.e. we are required to satisfy (**). If for given x , (**) does not have a solution it is an easy exercise to show that there exists $v \in X$ such that $v^te^{-tF}\ G = 0$ for all $t \geq 0$. If we successively differentiate this relation we get that $v^tF^q\ G = 0$ for all $q \geq 0$ i.e. rank $C < 0$, a contradiction which completes the proof.

Q.E.D.

Remark (2.4). The analysis above for continuous time linear systems applied essentially to the discrete time case $(T = \mathbb{Z})$ with one very important difference. Namely, given a discrete time system Σ of dimension n represented by difference equations

$$x(t+1) = F\,x(t) + G\,u(t)$$
$$y(t) = H\,y(t)$$

(F,G,H) matrices defined over an arbitrary field k of appropriate sizes, one can show that as in the continuous case Σ is completely reachable if and only if $\mathrm{rank}(G\ FG\ \ldots\ F^{n-1}G) = n$. However this simple criterion does not work for controllability. For this reason when we will be working over arbitrary fields (or commutative rings) we will use the term "completely reachable" instead of "completely controllable".

Example (2.5). We would like to give an example of a continuous system Σ which for physical reasons is obviously not completely controllable (or equivalently completely reachable) and then use the criterion of (2.3) to prove this rigorously.

A standard example (see e.g. [74]) is a system Σ in which the same force (or torque) u is applied to two rotating masses m_1 and m_2 which are not mechanically connected. Then if s_1 and s_2 are position variables the differential equations are

(*) $$m_i\,\ddot{s}_i = u \quad \text{for} \quad i = 1,2 .$$

To apply (2.3) we must put these equations in so-called state space form (see our discussion in Section 4). We use the elementary trick of introducing variables

$$x_1 = s_2\ ,\quad x_2 = \dot{s}_1\ ,\quad x_3 = s_2\ ,\quad x_4 = \dot{s}_2$$

and then the equations (*) are equivalent to the matrix equation

$$\dot{x} = Fx + Gu$$

where

$$F = \begin{bmatrix} 0 & 1 & 0 & 0 \\ 0 & 0 & 0 & 0 \\ 0 & 0 & 0 & 1 \\ 0 & 0 & 0 & 0 \end{bmatrix} \qquad G = \begin{bmatrix} 0 \\ 1/m_1 \\ 0 \\ 1/m_2 \end{bmatrix}$$

and $x^t = (x_1, x_2, x_3, x_4)$ (i.e. x is a column vector). It is easy to check that $\mathrm{rank}(G\ FG\ F^2G\ F^3G) < 4$ and hence Σ is not completely controllable.

§3. Constructibility and Observability

In this section we consider two concepts related to determining the state of a system from a knowledge of the availabke input/output data. As in Section 2, unless explicitly stated otherwise, Σ will denote a finite dimensional smooth linear dynamical system.

Remark (3.1). As noted in (1.8) with these assumptions, Σ may be represented as

$$\dot{x}(t) = F(t)x(t) + G(t)u(t)$$
$$y(t) = H(t)x(t) \ .$$

Let $x_0 := x(t_0)$. Then it is elementary to derive that

$$x(t) = M(t,t_0)\, x_0 + \int_{t_0}^{t} M(t,s)G(s)u(s)ds$$

where the matrix M satisfies the linear differential equation

$$\frac{\partial M(t,s)}{\partial t} = F(t)M(t,s) \quad , \quad s < t$$

$$M(s,s) = I \quad (I = \text{identity matrix}) \ .$$

M is usually referred to as the transition matrix associated with F .

Definitions-Remarks (3.2).

(i) An event (τ,x) (with respect to Σ) is unobservable if

$$H(s)\ M(s,\tau)x = 0 \quad \text{for all} \quad \tau < s \ .$$

Intuitively this means that an unobservable event at time τ cannot be detected from the outputs of a system after time τ . Σ is completely observable at time τ if no event (τ,x) of Σ is unobservable except $(\tau,0)$.

(ii) An event (τ,x) is unconstructible if $H(s)M(s,\tau)x = 0$ for all $s \leq \tau$. Intuitively this means that an event is unconstructible at time τ if it cannot be determined uniquely by the system output up to time τ . The definition of complete constructibility should be clear from (i).

Remark (3.3). Again one has notions of observability and constructibility in the non-linear case which we shall look at for polynomial systems in Part VI, Section 6 of these notes.

Analogously to (2.3) we have the following theorem:

Theorem (3.4). For Σ (with the above assumptions) constant and of dimension n , we have that the following statements are equivalent:

(i) Σ is completely constructible.

(ii) Σ is completely observable.

(iii) If $\Sigma = (F,G,H)$, then $\text{rank}(H^t\ F^t\ H^t\ \dots\ (F^t)^{n-1}\ H^t) = n$ where F^t and H^t are the transposed matrices.

Proof. Similar to that of (2.3). The details will be left as an exercise or see [82], pages 53-55.

Q.E.D.

Remarks (3.5).

(i) As for complete reachability, if Σ is a linear time-invariant finite dimensional discrete system, then Σ is completely observable if and only if condition (3.4)(iii) holds. This of course is true over arbitrary fields.

(ii) It should be clear that there is a duality between complete reachability and complete observability, and similarly between complete controllability and complete constructibility. Indeed it is easy to show for example that the pair of matrix functions $(F(t),H(t))$ define a completely observable system at time τ if and only if the pair of matrix functions $F^*(t) := F^t(2\tau-t)$, $G^*(t) := G^t(2\tau-t)$ define a completely reachable system at time τ. For details see [82].

Example (3.6). A standard example (see e.g. [74]) of an unobservable system is given by two rotating masses m_1 and m_2 which have their positions s_1 and s_2 controlled by separate torques u_1 and u_2 and the only observable output is given by a device which measures the difference $g = s_1-s_2$. Thus we have that $m_i\ddot{s}_i = u_i$ $i = 1,2$ and if we put this in state space form as in (2.5) we get that $\dot{x} = Fx+Gu$ where

$$F = \begin{bmatrix} 0 & 1 & 0 & 0 \\ 0 & 0 & 0 & 0 \\ 0 & 0 & 0 & 1 \\ 0 & 0 & 0 & 0 \end{bmatrix} \qquad G = \begin{bmatrix} 0 & 0 \\ 1/m_1 & 0 \\ 0 & 0 \\ 0 & 1/m_2 \end{bmatrix}$$

$u^t = (u_1,u_2)$, and x is as in (2.5).

The output is defined by $y = Hx$ where $H = [1\ 0\ {-1}\ 0]$. From (3.4) one sees that this system is not observable.

§4. State Variables

As indicated above, one of the fundamental techniques of modern system theory is the use of the state space form for linear smooth finite dimensional systems, i.e. one considers representations of the form

$$\dot{x} = Fx + Gu$$
$$y = Hx$$

where F, G, H may be time varying. For simplicity we will consider here free systems i.e. systems of the form $\dot{x} = Fx$.

It is a completely standard trick (which we have used in (2.5) and (3.6) above) that given a homogeneous differential equation of the n-th order

(1) $$\frac{d^n y}{dt^n} + \alpha_1 \frac{d^{n-1} y}{dt^{n-1}} + \ldots + \alpha_n y = 0$$

where the α_i are functions of t, (1) is equivalent to a system of n first order equations

$$\frac{dx_1}{dt} = x_2$$

(2) $$\frac{dx_2}{dt} = x_3$$

$$\vdots$$

$$\frac{dx_n}{dt} = -\alpha_n x_1 - \dots - \alpha_1 x_n$$

i.e. to $\dot{x} = Fx$ where

(3) $$F = \begin{bmatrix} 0 & 1 & 0 \dots & 0 \\ 0 & 0 & 1 \dots & 0 \\ \cdot & \cdot & \cdot \quad \cdot \quad \cdot & \cdot \\ 0 & 0 & 0 \dots & 1 \\ -\alpha_n & -\alpha_{n-1} & \dots & -\alpha_1 \end{bmatrix}$$

i.e. F is in rational canonical form.

In this section we would like to consider the problem of a given system of equations of the form $\dot{x} = \tilde{F}x$ ($\tilde{F}$ may be time varying), when we can write an equivalent homogeneous n-th order equation of the form (1). For example, if $\tilde{F}$ is time-invariant, this problem is completely trivial. Indeed in this case one explicitly solves $\dot{x} = \tilde{F}x$ by $x(t) = e^{\tilde{F}t} x(0)$. Clearly one gets an equivalent system by change of basis in the state space (which has the effect of transforming $\tilde{F} \longrightarrow g\tilde{F}g^{-1}$ where $g \in GL(n,k)$, $k = \mathbb{R}$ or $\mathbb{C}$, if $\tilde{F}$ is $n \times n$). Hence if $\tilde{F}$ is nonderogatory (i.e. admits a cyclic vector) by well-known results in linear algebra we may transform $\tilde{F}$ to rational canonical form F by change of basis in the state space and thus construct a homogeneous equation (1) which will be equivalent to the original system.

We remark that for $\tilde{F}$ nonderogatory, if $\tilde{g}$ is the corresponding cyclic vector then $(\tilde{F},\tilde{g})$ is a completely reachable pair. If we transform $\tilde{F}$ to rational canonical form F , $\tilde{g}$ transforms to

$$g = \begin{pmatrix} 0 \\ 0 \\ \vdots \\ 0 \\ 1 \end{pmatrix}$$

The canonical form (F,g) is referred to in the system theory literature as the control canonical form. See also our discussion in Part V, Section 2.

Following Deligne [26] we would like to investigate $\dot{x} = \tilde{F}x$ when $\tilde{F}$ is time-varying. This is a very beautiful argument and quite interesting since it contains several system-theoretic elements. Deligne considers in point of fact differential equations on a smooth connected complex manifold of dimension 1 , S , i.e. on a Riemann surface. (This is well-known [47] to be equivalent to a smooth irreducible algebraic curve.) The reader unfamiliar with Riemann surfaces may take S to be $\mathbb{C}$ (the complex numbers) or $\mathbb{C}\mathbb{P}^1$ (the Riemann sphere: $= \mathbb{C} \cup \{\infty\}$) , and the line bundle L below can be taken to be $\mathcal{O}_S$: = sheaf of holomorphic functions on S .

We will need to give some basic definitions from complex manifold theory (which actually has gained some popularity in system theory, see e.g. [73] or [145]). For complete details see the nice monograph of Chern [24] as well as Deligne [26].

Some Differential Geometry (4.1).

(i) Let M be a connected complex manifold and V a holomorphic vector bundle on M of rank n . Let Ω_M^1 : = holomorphic cotangent bundle on M . Then a connection on V is an operator $\nabla: V \to V \otimes \Omega_M^1$ and that ∇ is linear and for f a holomorphic function on an open subset $U \subset M$, and s a section over U , $\nabla(fs) = df \otimes s + f\nabla s$. (Recall that a section s over U is a holomorphic mapping $s: U \to V$ such that if $\pi: V \to M$ is the bundle morphism then $\pi \circ s$ = identity on U .)

(ii) Let ζ be a holomorphic vector field on M . Then one constructs from ∇ , a covariant differential operator $\nabla_\zeta: V \to V$ by setting $\nabla_\zeta(s) := \langle \nabla s, \zeta \rangle$ for s a local section (i.e. defined over some open subset $U \subset M$) . One also says that we contract ∇s along the vector field ζ .

(iii) A local section s is called horizontal if $\nabla s = 0$. From such sections we get local differential equations as follows: Over U a sufficiently small open subset of M choose $s_1,\ldots,s_n$ independent sections such that $s_1(x),\ldots,s_n(x)$ generate the fiber $V_x := \pi^{-1}(x)$ at each $x \in U$. Then we may write $\nabla s_i = \Sigma\, \alpha_j^i\, s_j$ where the α_j^i are holomorphic 1-forms. But for s any section over U , $s = \sum_i f_i s_i$ where the f_i are holomorphic functions on U , and thus $\nabla s = \sum_i (df_i + \sum_j f_j\, \alpha_j^i) s_i$. Hence if $\nabla s = 0$, we have that $df_i + \sum_j f_j\, \alpha_j^i = 0$ for $i = 1,\ldots,n$.

(iv) Let ζ , η be holomorphic vector fields. Then the curvature $R(\zeta,\eta) := \nabla_{[\zeta,\eta]} - [\nabla_\zeta,\nabla_\eta]$ where $[\zeta,\eta]$ is the Poisson bracket ([26]) and $[\nabla_\zeta,\nabla_\eta] := \nabla_\zeta \circ \nabla_\eta - \nabla_\eta \circ \nabla_\zeta$. ∇ is said to be integrable if $R = 0$.

(v) A local system $\tilde{V}$ on M is a sheaf of complex vector spaces which is locally isomorphic to one of the constant sheaves $\mathbb{C}^n$. It is easy to see that $V := \mathcal{O}_M \otimes_{\mathbb{C}} \tilde{V}$ is a vector bundle. (From Part I, Section 5, we can identify locally free sheaves and vector bundles.) Then one has the following essential fact proven in [26], page 12: V has a unique connection called the canonical connection for which the horizontal sections of V are the local sections of $\tilde{V}$. This connection moreover is integrable.

(vi) Given two holomorphic mappings $f: \mathbb{C} \to M$, $g: \mathbb{C} \to M$ which satisfy the condition that

$$\frac{\partial^j f_i}{\partial z^j}(0) = \frac{\partial^j g_i}{\partial z^j}(0) \quad , \quad i = 1,\ldots,m \ ; \quad j = 0,\ldots,r$$

where $f = (f_1,\ldots,f_m)$ and $g = (g_1,\ldots,g_m)$ locally on M (we assume dimension $M = m$) , we say that f and g are <u>r-equivalent.</u> Each equivalence class is called an <u>r-jet</u> and the set of all r-jets may be given a natural structure as a holomorphic vector bundle over M which we will denote by $T_r(M)$ (see e.g. Yano-Ishihara [153]). Note for example that $T_1(M) \cong T(M)$ the tangent bundle.

(vii) We now restrict ourselves to the case $M = S$ an arbitrary Riemann surface with local coordinate z . Denote by $J_r(S) := T_r(S)^*$ the dual bundle, so e.g. $J_1(S) = \Omega^1_S$ the cotangent bundle. We will regard $J_r(S)$ as a locally-free sheaf of $\mathcal{O}_S$-modules as in Part I, Section 5. Note that $J_r(S)$ contains $\Omega^r := (\Omega^1_S)^{\otimes r}$ as an $\mathcal{O}_S$-submodule. Moreover for a line bundle L (regarded as a locally free sheaf of rank 1), we set $J_r(L) := J_r(S) \otimes L$ and

With these standard definitions out of the way, we can now begin our main discussion. To motivate what we are going to do next, we give the following example:

<u>Example (4.2).</u> Let $S = \mathbb{C}$ with coordinate z . Then $J_r(\mathbb{C}) \cong \mathcal{O}^{r+1}_{\mathbb{C}}$ (direct sum of $\mathcal{O}_{\mathbb{C}}$ taken $r+1$ times). One then has a differential operator

$$D^r: \mathcal{O}_{\mathbb{C}} \to J_r(\mathbb{C}) \quad \text{given by}$$

$$f \longrightarrow (f, \frac{df}{dz}, \ldots, \frac{d^r f}{dz^r}) .$$

Consider an n-th-order homogeneous differential equation

$$(1') \qquad \frac{d^n y}{dz^n} + \alpha_1 \frac{d^{n-1} y}{dz^{n-1}} + \ldots + \alpha_n y = 0$$

where the α_i are holomorphic functions. This is essentially the same equation as (1) except with the stronger condition that the α_i are holomorphic. Now an equation of type (1') defines via its coefficients $\alpha_0(z),\ldots,\alpha_n(z)$ $(\alpha_0(z) \equiv 1)$ an element

$$\varphi \in \mathrm{Hom}_{\mathcal{O}_{\mathbb{C}}}(\bigoplus_{j=0}^{n} \mathcal{O}_{\mathbb{C}}, \mathcal{O}_{\mathbb{C}}) \cong \mathrm{Hom}_{\mathcal{O}_{\mathbb{C}}}(J_n(\mathbb{C}), \mathcal{O}_{\mathbb{C}}) .$$

But then a solution of (1') is exactly a holomorphic function f such that $\varphi(D^n f) = 0$.

It is precisely this situation which is generalized in [26]:

<u>Definitions (4.3).</u>

(i) Let L be a line bundle in a Riemann surface S . A <u>differential equation</u>

of order n on S with respect to L is an element $\varphi \in \mathrm{Hom}_{\mathcal{O}_S}(J_n(L),\Omega^n(L))$ which is the identity on $\Omega^n(L)$ (see (4.1)(vii)).

(ii) Given a section $s \in L(U)$, $U \subset M$ a sufficiently small open subset, one gets in the natural way a section $D^n(s) \in J_n(L)(U)$ (just truncate the Taylor series after terms of order n). Thus we can define a $\mathbb{C}$-linear homomorphism $D^n: L \to J_n(L)$. A solution of φ is a local section s such that $\varphi(D^n s) = 0$.

Remark (4.4). At the beginning of this section we have seen how to pass from a homogeneous differential equation (1) to a free system defined by (2) and (3). In (4.3) we have formalized (1). Now we need an appropriate generalization of (2). The crucial point is that as we noted in the constant case, the matrix F admits a cyclic vector and it is precisely this property which allows us to reconstruct the differential equation. Thus it is this notion of cyclicity which we must generalize now:

Cyclic Sections (4.5).

(i) Let V be a holomorphic vector bundle of rank n over S, s a local section defined on $U \subset M$, ∇ a connection, ζ a holomorphic vector field, and ∇_ζ the associated covariant differentiation. Then s is cyclic if and only if the $(\nabla_\zeta)^i(s)$ generate V over U for $i = 0,\ldots,n-1$. We leave it as a simple exercise to show that this definition is independent of ζ.

(ii) Given $\mu: V \to L$ an $\mathcal{O}_S$-homomorphism with L a line bundle, μ is said to be cyclic, if μ considered as a section of $V^* \otimes L$ is locally cyclic where we locally identify L with $\mathcal{O}_S$. (Note one has that $\mathrm{Hom}_{\mathcal{O}_S}(V,L) \cong V^* \otimes L$; see [52], page 123.)

(iii) It is easy to see that $J_r(L)$ is a graded $\mathcal{O}_S$-module with the i-th graded piece being isomorphic to $\Omega^i(L)$ for $0 \leq i \leq r$. Consequently we have a natural $\mathcal{O}_S$-homomorphism $\lambda_r: J_r(L) \to L$.

We can now state the following fundamental theorem:

Theorem (4.6). (Cauchy-Deligne). A differential equation φ of order n with respect to a line bundle L over S defines a canonical connection on $J_{n-1}(L)$ for which the horizontal sections are the images by D^{n-1} of solutions of φ. Moreover the natural homomorphism $\lambda_{n-1}: J_{n-1}(L) \to L$ is cyclic.

Proof. For a proof of the first part of the theorem see [26] especially pages 12 and 25. Because of the system theoretic flavor and to illustrate the definitions, assuming the first part of the theorem we will show that λ_{n-1} is cyclic. The problem is local and so we are in the situation of (4.2) from which we use the same notation. Thus we have our equation φ of the form (1') given in (4.2). By definition s is horizontal if and only if $\nabla s = 0$. But from the first part of (4.6) s is horizontal if and only if there exists f a solution of (1') such that

$s = D^{n-1}f$. We consider s as a column vector with components s_i , $i = 1,\ldots,n$. Then it is easy to see the above discussion implies s is horizontal if and only if $\dot{s} = Fs$ where F is a matrix in rational canonical form (3) and that F is precisely the matrix of the connection ∇ . It is now trivial to verify that λ_{n-1} is cyclic.

Q.E.D.

State Space and Homogeneous Differential Equations (4.7).

(i) From our discussion above it should be obvious that the solutions of φ (notation as in (2.6)) are precisely the images under λ_{n-1} of the horizontal sections of $J_{n-1}(L)$. This is precisely the fact that we may pass from a differential equation to its state space form. In point of fact, the preceding discussion should make it clear that the correct formalization of the state space form $\dot{x} = Fx$ is a triple (V,L,λ) consisting of a rank n vector bundle V with a connection, a line bundle L , and a cyclic $\mathcal{O}_S$-homomorphism $\lambda: V \to L$.

Now to a pair (φ,L) consisting of a differential equation φ of order n with respect to a line bundle L , we have associated the triple $(J_{n-1}(L),L,\lambda_{n-1})$ satisfying these state space conditions. The point now is to see how to associate a pair (φ,L) to such a triple (V,L,λ) .

(ii) Given (V,L,λ) , the fact that λ is cyclic automatically means that V is isomorphic to $J_{n-1}(L)$. Indeed to define an $\mathcal{O}_S$-module homomorphism on V it is enough ([26], page 12) to define a $\mathbb{C}$-linear map on the local system of horizontal sections of V . Thus one can define a homomorphism $\alpha_r: V \to J_r(L)$ by setting for a horizontal section $s \in V$, $\alpha_r(s) = D^r(\lambda(s))$. A local check shows that α_{n-1} is an isomorphism (V has rank n).

(iii) Now it should be clear how to go from a triple (V,L,λ) as in (i) to a pair (φ,L) . Indeed recalling the graded structure on $J_r(L)$ ((2.5)(iii)) for $r \geq s$ we have a natural projection $\pi_{r,s}: J_r(L) \to J_s(L)$ such that $\pi_{r,s} \circ \alpha_r = \alpha_s$. But from (4.7)(iii) α_{n-1} is an isomorphism, and thus we see that locally $J_n(L)$ is the direct sum of $\alpha_n(V)$ and $\ker \pi_{n,n-1} \cong {}^n(L)$. But by definition a differential equation of order n with respect to L is just a homomorphism $\varphi: J_n(L) \to \Omega^n(L)$ which induces the identity on $\Omega^n(L)$. Hence there exists a unique such φ which sends $\alpha_n(V)$ to zero. Consequently, to (V, L ,λ) we associate (φ,L) .

Finally it is easy to check that if we apply the construction mentioned in (4.7)(i) to (φ,L) to get a state space triple, this triple will be isomorphic to (V, L ,λ) . The key fact needed is that if s is a horizontal section, then $\varphi(D^n\lambda(s)) = \varphi(\alpha_n(s)) = 0$, i.e. $\lambda(s)$ is a solution of φ . But from (2.7)(i) solutions of φ are precisely the images under λ_{n-1} of $J_{n-1}(L)$.

We summarize our discussion with the following theorem ([26], page 27):

Theorem (4.8). There exists a natural 1-1 correspondence between isomorphism classes of pairs (φ, L) where φ is an n-th order differential equation on a Riemann surface S with respect to a line bundle L, and isomorphism classes of triples (V, L, λ) where V is a rank n vector bundle with a connection on S and $\lambda: V \to L$ is a cyclic homomorphism.

Proof. Immediate from (4.7). We leave it as a simple exercise for the reader to define the proper notions of isomorphism for the objects given in (4.8).

Q.E.D.

5. Transfer Functions

In contrast to the high-powered techniques of Section 4, we would like to conclude our sketch of system theory with some definitions and remarks concerning a simple and modest (but very pretty) object: the system transfer function. In point of fact, transfer function techniques still pervade the engineering literature and have been seriously argued to be in many respects superior to the state space methods which we use in these notes. (For an interesting discussion of this see Horowitz-Shaked [68].) We also will make contact here with some topics which we shall consider in greater detail a bit later on in these notes.

Transfer Functions and their Realizations (5.1).

(i) Let $\dot{x} = Fx+Gu$, $y = Hx$ define a time-invariant system Σ over $\mathbb{R}$ of dimension n, with m inputs and p outputs. Suppose that the input functions $u(t) = (u_1(t), \ldots, u_n(t))$ are such that

$$\int_0^\infty |u_i(t)| e^{-\sigma t} \, dt < \infty$$

for some finite real σ. Assume moreover that $x(0) = 0$. Then one can take the Laplace transform of the system Σ. Recall e.g. that for the $u_i(t)$,

$$U_i(z) := \text{Laplace transform of } u_i(t) := \int_0^\infty u_i(t) e^{-zt} \, dt$$

and the conditions on the u_i mean that these integrals converge. For the properties of the Laplace transform from a control-theoretic point of view see [74], Chapter 1.

Let then $U(z)$, $X(z)$, $Y(z)$ be the Laplace transforms of $u(t)$, $x(t)$, $y(t)$ respectively. As is well known, the Laplace transform replaces differentiation with respect to t by multiplication by z. Consequently, since by assumption $x(0) = 0$ we have that the Laplace transform of Σ is

$$zX(z) = FX(z) + GU(z)$$
$$Y(z) = HX(z)$$

and hence

$$T(z) := \frac{Y(z)}{X(z)} = H(zI-F)^{-1} G$$

where I is the $n \times n$ identity matrix. $T(z)$ is a $p \times m$ matrix with entries consisting of real rational functions called the transfer function of the system Σ .

(ii) Even though we have derived the transfer function of a real system using analysis, the transfer function is in point of fact an algebraic object. Indeed given any constant system $\Sigma = (F,G,H)$ defined over any field k ((1.10), (iii)), we can define a transfer function for Σ by setting $T_\Sigma(z) := H(zI-F)^{-1} G$.

(iii) Let $\Sigma = (F,G,H)$ be as in (ii). We say that Σ is canonical if it is completely reachable and completely observable. We leave it as a simple exercise (just use Cramer's rule) for the reader to show that if Σ is canonical then the transfer function $T_\Sigma(z)$ is strictly proper i.e. all the entries of $T_\Sigma(z)$ are rational functions with the degree of the denominator greater than the degree of the numerator and with no common factors.

(iv) Now suppose we are given a strictly proper rational function $T(z)$ over a field k . In Part VI we will see that it is possible to canonically realize $T(z)$ by a canonical system $\Sigma = (F,G,H)$, i.e. Σ has the property that $H(zI-F)^{-1}G = T(z)$. Moreover we will show that this canonical realization is unique up to change of basis in the state space k^n , i.e. if $\tilde{\Sigma} = (\tilde{F}, \tilde{G}, \tilde{H})$ is another canonical realization then there exists $g \in GL(n,k)$ (the group of invertible $n \times n$ matrices) such that $\tilde{F} = gFg^{-1}$, $\tilde{G} = gG$, $\tilde{H} = Hg^{-1}$. In particular transfer functions give a coordinate-free description of canonical systems.

(v) The discussion in (iv) will be one of our fundamental motivations in Part IV of these notes to study the orbit space of the set of triples of matrices $\{(F,G,H)\} \cong k^{n^2+nm+np}$ (where F is $n \times n$, G is $n \times m$, H is $p \times n$) under the action of $GL(n,k)$ given in (iv). For k algebraically closed we will use Mumford's geometric invariant theory to study this space.

(vi) In case that $\Sigma = (F,G,H)$ is a canonical scalar input/output system $(m = p = 1)$, the transfer function $T(z)$ is just a strictly proper rational function which we can view as a coordinate-free model of a linear time-invariant system. Now it is easy to check that if $\dim \Sigma = n$, then the degree of the denominator of $T(z)$, i.e. the McMillan degree of $T(z)$, is n . Motivated by identification theory, we will study the topology of the space of all real or complex transfer functions of McMillan degree n in Part VII.

Finally we conclude this section with a slightly different perspective on transfer functions:

Input/Output Maps (5.2).

(i) Let Σ be a finite dimensional linear smooth time-varying system defined by

$$\dot{x}(t) = F(t)x(t)+G(t)u(t)$$
$$y(t) = H(t)x(t) .$$

Set $x_0 := x(t_0)$. Recall in (3.1) we found that

$$x(t) = M(t,t_0)x_0 + \int_{t_0}^{t} M(t,s)G(s)u(s)ds$$

where M is the transition matrix associated with F . We suppose that $x_0 = 0$. Then we see that

$$y(t) = \int_{t_0}^{t} H(t)M(t,s)G(s)u(s)ds .$$

The function $H(t)M(t,s)G(s)$ is called the <u>impulse response function</u> since it measures the output due to a delta function.

(ii) In case Σ is time-invariant, and we assume $x_0 = 0$, we get that

$$x(t) = \int_{t_0}^{t} e^{F(t-s)}Gu(s)ds$$

and thus

$$y(t) = \int_{t_0}^{t} He^{F(t-s)}Gu(s)ds .$$

From the input/output point of view we may regard the system Σ as associating to the input $u(t)$ the output $y(t)$. Define $f_\Sigma u(t) := y(t)$ i.e. f_Σ is an input/output map (see the discussion in Section 1). Note moreover if $\dim \Sigma = n$, and if we set $\Sigma^g := (gFg^{-1}, gG, Hg^{-1})$ where $g \in GL(n,k), k = \mathbb{R}$ or $\mathbb{C}$, then $f_\Sigma = f_{\Sigma^g}$, i.e. the input/output behavior is invariant under change of basis in the state space or in other words f_Σ gives a coordinate free description of the system. Thus in the canonical case from (5.1)(iv) we see that we may identify input/output maps with transfer functions.

In the discrete time case over an arbitrary field k when Σ is represented by difference equations (see (1.10)(ii)) one has a similar description as that given here for the continuous time case in terms of an input/output map and a $GL(n,k)$ action. We leave the details to the reader or see [82].

(iii) Note again, and this is crucial, it is precisely because of the arbitrary choice of coordinates in describing a finite dimensional constant system of dimension n in state space form that one is forced to go to the $GL(n,k)$-quotient space of the space of all such systems (with fixed input and output dimensions) in order to describe the input/output behaviors. This redundancy inherent in the powerful state space description and the ensuing quotient spaces (which we will see can be regarded as moduli spaces of linear systems) will be the topic of study in Part IV.

(iv) We make contact once more with realization theory. Using the notation of (ii) if one takes the inputs for the system Σ to be delta functions it is easy to see that one gets outputs of the form $G, HFG, HF^2G, \ldots$. Thus from an input/output point of view, we can regard f_Σ as a sequence of $p \times m$ matrices $\{A_i\}_{i\geq 1}$ (where H is $p \times n$ and G is $n \times m$). Conversely, given such a sequence of matrices $\{A_i\}_{i\geq 1}$ we can ask when we can realize this sequence, i.e. when does there exist (F,G,H) such that $A_i = H F^{i-1} G$ for all $i \geq 1$? Moreover we would like to characterize all such realizations and if possible find one canonical in some sense. Note again these questions make sense over any field k or over any commutative ring with unity R . This will be one of the major themes in Part VI.

PART III. INVARIANT THEORY AND ORBIT SPACE PROBLEMS

As alluded to in Section 5 of Part II, an essential aspect of these notes will be the study of certain quotients of affine varieties by algebraic groups which come up in system theory. These quotients are very closely connected to the theory of moduli.

In his spectacular monograph [104] David Mumford modernized classical invariant theory and created a new subject "geometric invariant theory" which has become the essential took for studying global moduli problems in algebraic geometry. In his talks given at Oslo [108] this theory was applied to the case of the moduli of endomorphisms of finite dimensional vector spaces and as we will see in Part IV one needs only a small generalization of this theory in order to describe the quotients (in particular the Kalman space of completely reachable pairs [79]) and the global moduli of linear dynamical systems.

Thus in this part of these lecture notes after sketching the basic theory of algebraic groups we will discuss in some detail geometric invariant theory, especially from [108].

Throughout Part III, k will denote a fixed algebraically closed field.

§1. Algebraic Groups

In this section we will sketch that part of the theory of algebraic groups which we will need in the sequel. For further details see Borel [10] and Humphreys [71].

Definition (1.1). Let G be a variety endowed with the structure of a group. If $u: G \times G \to G$, $u(x,y) = xy$, and $i: G \to G$, $i(x) = x^{-1}$ are morphisms, then G is called an algebraic group.

Exercise. Prove that a closed subgroup of an algebraic group is again an algebraic group.

Examples (1.2).

(i) Let $M_{n,n}(k)$ be the group of $n \times n$ matrices over k . Then it is clear that $M_{n,n}(k)$ is an algebraic group (under addition). If we let $n=1$, we denote $\mathbb{G}_a := M_{1,1}(k)$ which is of course just $\mathbb{A}^1$ given a group structure via translation.

(ii) Let $GL(n,k) = \{A \in M_{n,n}(k) \mid \det A \neq 0\}$. Then it is clear that $GL(n,k)$ is a multiplicative algebraic group and being the complement of a hypersurface in affine space must also be affine. We denote $G_m := GL(1,k)$.

(iii) From the exercise, if $T(n,k) \subset GL(n,k)$ is the subgroup of upper triangular matrices, $D(n,k) \subset GL(n,k)$ the diagonal subgroup, and $U(n,k) \subset GL(n,k)$ the unipotent subgroup (upper triangular matrices with all diagonal entries 1), all these groups are algebraic groups.

(iv) Let $SL(n,k) = \{A \in M_{n,n}(k) \mid \det A^{-1} = 0\}$. Then $SL(n,k)$ (the "special linear group") is clearly an algebraic group which as a variety is the hypersurface $\det A-1 = 0$ for $A \in M_{n,n}(k)$. Therefore $\dim SL(n,k) = n^2-1$.

Exercise. Prove $U(2,k) \cong \mathbb{G}_a$.

We will be especially interested in algebraic groups of the following kind:

Definition (1.3). A linear algebraic group is an algebraic group which is isomorphic to a closed subgroup of $GL(n,k)$.

The importance of this definition should be indicated by:

Theorem (1.4). Any affine algebraic group is linear.

Proof. See [71], page 63.

Q.E.D.

Remark (1.5). Theorem (1.4) means that any affine algebraic group may be realized as a group of matrices. This fact is an enormous technical aid in many proofs involving affine algebraic groups.

We will be interested in the sequel in studying actions of algebraic groups on varieties. By "action" we mean:

Definition (1.6). Let G be an algebraic group and X a variety. Then an action of G on X is a morphism $\varphi: G \times X \to X$ which satisfies the standard properties of actions of groups on sets.

We can now state the following fundamental theorem:

Theorem (1.7). ("Closed orbit lemma"). Let G be an algebraic group acting on a variety X. Then every orbit is a smooth locally closed subset of X whose boundary is a union of orbits of strictly lower dimension.

Proof. We give the main idea here. Complete proofs may be found in both [10] and [71].

Let $x \in X$. Define $u: G \to X$ by $g \longmapsto g \cdot x$. Then u is a morphism whose image is the orbit of x, $0(x)$. But by the Chevalley constructibility theorem (see Section 4 of Part I), $0(x)$ is constructible. Using the fact that G acts transitively on $0(x)$ and stablizes $\overline{0(x)}$ it is then immediate that $0(x)$ is open in $\overline{0(x)}$, and thus the boundary $\overline{0(x)} - 0(x)$ is closed. The boundary is stable under G and hence must be a union of orbits of strictly lower dimension.

Q.E.D.

Remarks (1.8).

(i) (1.7) implies that orbits of minimal dimension are closed.

(ii) Recall that $(F,G) \in M_{n,n}(k) \times M_{n,m}(k)$ is completely reachable if

$\mathrm{rank}(G\ FG \dots F^{n-1}G) = n$. In Part IV of these notes we will show that (F,G) is completely reachable if and only if $\dim \mathrm{stab}(F,G) = 0$. Let $V_{n,m} = \{(F,G) \text{ completely reachable}\}$. Then $V_{n,m}$ is a Zariski open subset of $M_{n,n}(k) \times M_{n,m}(k)$. Moreover under the action of $GL(n,k)$ on $M_{n,n}(k) \times M_{n,m}(k)$ by $(F,G) \longmapsto (gFg^{-1}, gG)$ for $g \in GL(n,k)$, it is clear that $V_{n,m}$ is invariant.

Now since the dimension of the stabilizers is constant on $V_{n,m}$, it is an immediate consequence of the closed orbit lemma (1.7) that $GL(n,k)$ acts on $V_{n,m}$ with closed orbits. This remark will be crucial to the results of Part IV.

§2. On the Moduli of Endomorphisms

In this section we consider a classical moduli problem, the problem of classifying endomorphisms of vector spaces of fixed dimension. The problem of finding the "moduli" or "parameters" upon which a given class of mathematical objects depends is very old and at least goes back to Riemann. Indeed, Riemann claimed that the general Riemann surface of genus g depended on "3g-3 moduli". Our discussion below is almost completely based on David Mumford's Oslo talks [108] in which he attempts to give an elementary exposition of geometric invariant theory and the corresponding problems of moduli. We will only give that part of his treatment which will be essential for us in Part IV of these notes. The reader is therefore urged to see the complete talks in [108].

In general, a "moduli-problem" consists of the classification of a set of algebraic objects. For our restricted purposes we will require this classification to have two parts:

(i) Find the equivalence classes of algebraic objects under the relationship of isomorphism.

(ii) Parametrize the equivalence classes by some algebraic object, most preferably an algebraic variety.

The moduli problem we want to consider here is the classification of pairs (V,L) where V is an n-dimensional vector space over $k = \bar{k}$ and L is an endomorphism of V . In this case the answer to (i) is classical, namely the Jordan canonical form. That is, we can always find a basis for V such that L may be represented by the Jordan canonical form. Thus to solve problem (ii) (if a solution exists?) we must somehow parametrize Jordan canonical forms or equivalently families of endomorphisms. This leads to the following definition:

Definitions (2.1).

(i) Let S be any k-variety. Then a family of endomorphisms of dimension n over S is a pair $(\mathcal{D}, \mathcal{L})$ where $\mathcal{D}$ is a rank n vector bundle over S

and and $\mathcal{L}$ is a vector bundle endomorphism.

(ii) For each S a k-variety, let $\mathcal{F}(S) := \{$isomorphism classes of families of endomorphisms of dimension $n\}$.

Remark (2.2). Let $Var(k) :=$ category of varieties over k, Sets: $=$ category of sets. Given S, $S' \in Var(k)$, and a morphism $\varphi: S \to S'$, if $(\mathcal{V},\mathcal{L})$ is a family of endomorphisms of dimension n over S', then we can pull $(\mathcal{V},\mathcal{L})$ back to S via $(\varphi^*\mathcal{V}, \varphi^*\mathcal{L})$. Note this means that if $\varphi^*\mathcal{V}(s)$ denotes the fiber of $\varphi^*\mathcal{V}$ over s, then we can identify $\varphi^*\mathcal{V}(s)$ with $\mathcal{V}(\varphi(s))$. $\varphi^*\mathcal{L}$ is defined similarly.

Consequently the correspondence $S \longmapsto \mathcal{F}(S)$ is a contravariant functor from $Var(k)$ to Sets. Since we are interested in parametrizing families of endomorphisms, the "correct" definition of "moduli space" should be made in terms of the representability properties of the functor $\mathcal{F}$. More precisely we have the following crucial definitions:

Definitions (2.3).

(i) Let $\mathcal{M}$ be any k-variety. Then $h_{\mathcal{M}}$ denotes the contravariant functor from $Var(k)$ to Sets defined by $h_{\mathcal{M}}(S) := Hom(S,\mathcal{M})$ (where $Hom(S,\mathcal{M}) :=$ set of morphisms from S to $\mathcal{M}$).

(ii) A morphism $\varphi: \mathcal{F} \to h_{\mathcal{M}}$ is a correspondence which to each $S \in Var(k)$ assigns a mapping $\varphi(S): \mathcal{F}(S) \to Hom(S,\mathcal{M})$ of sets.

(iii) A coarse moduli space for families of endomorphisms of dimension n is a pair $(\mathcal{M},\varphi)$ consisting of a variety $\mathcal{M}$ and a morphism $\varphi: \mathcal{F} \to h_{\mathcal{M}}$ such that:

(a) for p a point (i.e. $p = Spec\ k$), $\varphi(p): \mathcal{F}(p) \to Hom(p,\mathcal{M}) \cong \mathcal{M}$ is bijective;

(b) for any $N \in Var(k)$ and for any morphism $\psi: \mathcal{F} \to h_N$, there exists a unique morphism $f: \mathcal{M} \to N$ such that the natural diagram

$$\begin{array}{ccc} & & Hom(S,\mathcal{M}) \\ & \overset{\varphi(S)}{\nearrow} & \\ \mathcal{F}(S) & & \downarrow Hom(S,f) \\ & \underset{\psi(S)}{\searrow} & \\ & & Hom(S,N) \end{array}$$

commutes for every $S \in Var(k)$.

(iv) A fine moduli space for families of endomorphisms of dimension n is a pair $(\mathcal{M},\varphi)$ consisting of a variety $\mathcal{M}$ and a morphism $\varphi: \mathcal{F} \to h_{\mathcal{M}}$ such that for every $S \in Var(k)$, $\varphi(S): \mathcal{F}(S) \to Hom(S,\mathcal{M})$ is a bijection. $\mathcal{M}$ is also said to represent the functor $\mathcal{F}$.

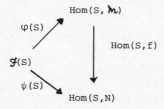

Remarks (2.4).

(i) It is easy to see that a fine moduli space is a coarse moduli space, but the converse may not be true.

(ii) From property (a) of the definition of coarse moduli space, we have that the points of $\mathcal{M}$ are in 1-1 correspondence with isomorphism classes of pairs (V,L) as above or equivalently with Jordan canonical forms. From property (b) it is an easy exercise to show if a coarse moduli space exists, then it must be unique.

(iii) There is nothing special about the endomorphism functor $\mathcal{F}$, and one can apply the definitions (2.3) to any contravariant functor from $\mathrm{Var}(k)$ to Sets in order to get a notion of moduli space. This will be done in Part IV.

We have discussed a general theory of moduli spaces, applied it to $\mathcal{F}$ and now the only problem is to show a moduli space exists. Unfortunately, despite the previous fancy mathematical dressing for the Jordan canonical form, for $n \geq 2$ a moduli space never exists (not even a coarse one)! Indeed we have the following example:

Example (2.5). We consider the case $n = 2$. Suppose that $\mathcal{M}$ is a coarse moduli space for $\mathcal{F}$. Let $(\mathcal{V}, \mathcal{L})$ be the family of endomorphisms of dimension 2 over the line $\mathbb{A}^1$ defined by letting $\mathcal{V} = \mathbb{A}^1 \times k^2$ (the trivial bundle) and

$$\mathcal{L} = \begin{pmatrix} 1 & t \\ 0 & 1 \end{pmatrix}$$

where $t \in \mathbb{A}^1$ is a coordinate. Since $\mathcal{M}$ is a coarse moduli space for $\mathcal{F}$, we have a mapping $\varphi(\mathbb{A}^1)\colon \mathcal{F}(\mathbb{A}^1) \to \mathrm{Hom}(\mathbb{A}^1, \mathcal{M})$, and so let $f\colon \mathbb{A}^1 \to \mathcal{M}$ be a morphism corresponding to $(\mathcal{V}, \mathcal{L})$.

Now note that for $t \neq 0$ all the fibers of $(\mathcal{V}, \mathcal{L})$ over $\mathbb{A}^1$ are isomorphic (i.e. the Jordan canonical form for $\begin{pmatrix} 1 & t \\ 0 & 1 \end{pmatrix}$, $t \neq 0$ is $\begin{pmatrix} 1 & 1 \\ 0 & 1 \end{pmatrix}$). This means that f is constant on $\mathbb{A}^1 - \{0\}$. By continuity then f must be constant on all of $\mathbb{A}^1$. But the special fiber at $t = 0$ of $(\mathcal{V}, \mathcal{L})$ gives $\begin{pmatrix} 1 & 0 \\ 0 & 1 \end{pmatrix}$ which is not isomorphic to $\begin{pmatrix} 1 & t \\ 0 & 1 \end{pmatrix}$, $t \neq 0$ (the Jordan canonical forms are different). Consequently $\mathcal{M}$ cannot exist because of this "jump". In Part V we will discuss a remedy for this discontinuity in the Jordan canonical form.

Remark (2.6). When we discuss quotients below we will give the precise reason for the failure of the existence of a moduli space for $\mathcal{M}$ (it is closely related to the closedness of the orbits).

We can give here though a heuristic reason which will help us find a subfunctor of $\mathcal{F}$ which can be represented. Consider the Jordan canonical form of $L\colon V \to V$,

$$L_* = \begin{pmatrix} J_1 & & & 0 \\ & J_2 & & \\ & & \ddots & \\ 0 & & & J_r \end{pmatrix}$$

where each of the J_i is a Jordan block consisting of one of the eigenvalues of L on the diagonal and 1's on the superdiagonal. Thus L_* has the eigenvalues of L on the diagonal and 1's or 0's on the superdiagonal.

From Example (2.5) it should be clear that the continuous moduli of (V,L) are the eigenvalues on the diagonal and hence if there is any chance at all of finding a moduli space we must consider those pairs (V,L) with a fixed pattern of 1's and 0's on the superdiagonal of the Jordan canonical form. There are two obvious possibilities:

(i) all 0's , i.e. the semi-simple case;

(ii) all 1's , the case in which L has a cyclic basis vector v ,

i.e. a vector s.t. $v, Lv, \ldots, L^{n-1}v$ span V . Note that (L,v) is then a completely reachable pair.

Now in [108] Mumford considers both cases (i) and (ii) and ahows for (i) that $\mathbb{A}^n$ is a coarse but not a fine moduli space for semisimple endomorphisms when $n > 1$. However for us (ii) is clearly the more important case, and thus this is the case we will consider now.

We begin with the following proposition:

Proposition (2.7). Let $\mathfrak{F}$ be as above. Then there exists a morphism of functors $\mathfrak{F} \xrightarrow{u} h_{\mathbb{A}^n}$ such that $\mathfrak{F}(\text{point}) \to \mathbb{A}^n$ is given by $(V,L) \longmapsto (c_1(L), \ldots, c_n(L))$ where the $c_i(L)$ are the characteristic coefficients of L .

Proof. Mumford [108], page 174 gives a proof using algebraic-geometric methods. When we have stated Richardson's criterion in Section 5 we will give an alternate proof to which the reader is referred.

Q.E.D.

Definition (2.8). Let $\mathfrak{F}_{cr}$ ("cr" stands for "completely reachable") be the subfunctor of $\mathfrak{F}$ defined by

$\mathfrak{F}_{cr}(S) = \{(\mathcal{V}, \mathcal{L}) \in \mathfrak{F}(S)$ such that $\mathcal{V}$ has a cyclic section with respect to $\mathcal{L}$, i.e. there exists a section $\sigma: S \to \mathcal{V}$ such that $\sigma, \mathcal{L}\sigma, \ldots, \mathcal{L}^{n-1}\sigma$ span $\mathcal{V}\}$.

Theorem (2.9). Let $u': \mathfrak{F}_{cr} \to h_{\mathbb{A}^n}$ be the restriction of the morphism $u: \mathfrak{F} \to h_{\mathbb{A}^n}$ of (2.7) to $\mathfrak{F}_{cr}$. Then u' is an isomorphism of functors, i.e. $\mathbb{A}^n$ is a fine moduli space for $\mathfrak{F}_{cr}$.

Proof. The proof should be obvious to the system theorist. Indeed we need only use the "sharp canonical form". More precisely let $(\mathcal{V},\mathcal{L}) \in \mathcal{F}_{cr}(S)$. Then if σ is a cyclic section, $\mathcal{V}$ is free with basis $\sigma, \mathcal{L}\sigma, \ldots, \mathcal{L}^{n-1}\sigma$ and the matrix of $\mathcal{L}$ with respect to this basis is

$$\begin{bmatrix} 0 & 0 & \ldots & -c_n \\ 1 & 0 & \ldots & -c_{n-1} \\ 0 & 1 & \ldots & -c_{n-2} \\ \vdots & \vdots & & \vdots \\ 0 & 0 & & -c_1 \end{bmatrix}$$

where the c_i are the characteristic coefficients. Thus elements of $\mathcal{F}_{cr}(S)$ are in 1-1 correspondence with n-tuples $\mathcal{O}(S)^n$ i.e. with morphisms $S \to \mathbb{A}^n$.

Q.E.D.

Remark (2.10). We will see in Part IV that (2.9) is the template on which all the moduli space proofs (see e.g. [53], [54], [57]) of linear system theory are built.

§3. Quotients

The notation and conventions of Section 2 are again in force here.

Remark (3.1). Let $\mathcal{V} := \mathbb{A}^{n^2} \times k^n$ be the trivial vector bundle over $\mathbb{A}^{n^2}$. The ring of regular functions on $\mathbb{A}^{n^2}$ is of course $k[t_{ij}]_{\substack{1\leq i\leq n \\ 1\leq j\leq n}}$. Let $\mathcal{L} = (t_{ij})_{\substack{1\leq i\leq n \\ 1\leq j\leq n}}$. Then the matrix $\mathcal{L}$ defines an endomorphism of $\mathcal{V}$.

Now the pair $(\mathcal{V},\mathcal{L})$ has the essential property that given a variety S and any family of endomorphisms of dimension n over S $(\mathcal{V}',\mathcal{L}')$, $(\mathcal{V}',\mathcal{L}')$ is locally induced by $(\mathcal{V},\mathcal{L})$. More precisely, let $\{U_\alpha\}$ be an open covering of S such that $\mathcal{V}'|U_\alpha$ is trivial $(\cong U_\alpha \times \mathbb{A}^n)$ for each U_α . Then $\mathcal{L}'|U_\alpha$ may be written in matrix form relative to a basis of $\mathcal{V}'|U_\alpha$ and the entries of this matrix will define a morphism $f_\alpha: U_\alpha \to \mathbb{A}^{n^2}$. It is then immediate to check that $(\mathcal{V}'|U_\alpha, \mathcal{L}'|U_\alpha) \cong (f_\alpha^*\mathcal{V}, f_\alpha^*\mathcal{L})$. Consequently if $\varphi: \mathcal{F} \to h_{\mathcal{M}}$ is any morphism of functors, the properties of φ are uniquely determined by the morphism $\alpha: \mathbb{A}^{n^2} \to \mathcal{M}$ associated to $(\mathcal{V},\mathcal{L})$.

Next we can regard $\mathbb{A}^{n^2}$ as $M_{n,n}(k)$ the set of $n \times n$ matrices, and $GL(n,k)$ acts on $M_{n,n}(k)$ by conjugation. Given $A \in M_{n,n}(k)$, it is trivial to see that the fibers of $(\mathcal{V},\mathcal{L})$ over A and gAg^{-1} are isomorphic, i.e. α is constant on the orbits.

Conversely it is easy to check that if $\alpha: \mathbb{A}^{n^2} \to \mathcal{M}$ is a morphism constant on the $GL(n,k)$ orbits, then there exists a morphism of functors $\varphi: \mathcal{F} \to h_{\mathcal{M}}$ which associates α with $(\mathcal{V},\mathcal{L})$ on $\mathbb{A}^{n^2}$. We have thus proven the following crucial theorem:

Theorem (3.2). There exists a canonical 1-1 correspondence between morphisms $\varphi: \mathcal{F} \to h_{\mathfrak{M}}$ and morphisms $\alpha: \mathbb{A}^{n^2} \to \mathfrak{M}$ constant on the $GL(n,k)$ orbits.

Then the property (b) of (2.3)(iii) in the definition of coarse moduli space translates via (3.2) into the following notion of "quotient":

Definition (3.3). Let G be an algebraic group acting on a variety X. A quotient of X by G is a pair (Y,α) where Y is a variety and $\alpha: X \to Y$ is a morphism such that

(i) α is constant on the orbits;

(ii) given Y' a variety, $\alpha': X \to Y'$ a morphism constant on the orbits, there exists a unique morphism $\beta: Y \to Y'$ such that the diagram

$$\begin{array}{ccc} X & \xrightarrow{\alpha} & Y \\ & \underset{\alpha'}{\searrow} & \downarrow \beta \\ & & Y' \end{array}$$

commutes. Note a quotient, if it exists, is unique up to isomorphism.

Example (3.4). Consider the morphism $\alpha: A^{n^2} \to A^n$ defined by sending an $n \times n$ matrix A to $(c_1(A),\dots,c_n(A))$ where the $c_i(A)$ are the characteristic coefficients. Using Richardson's criterion in Section 5 we will show that $(\mathbb{A}^n,\alpha)$ is a quotient for $\mathbb{A}^{n^2}$ relative to $GL(n,k)$ acting on $\mathbb{A}^{n^2}$ by conjugation. Note that from Theorem (3.2), this will give an invariant-theoretic proof of Proposition (2.7) above.

Remark (3.5). We can now explain precisely why a coarse moduli space does not exist for the functor $\mathcal{F}$ of families of endomorphisms. Indeed from our above discussion if a coarse moduli space existed it must be $\mathbb{A}^n$.

But the coarse moduli space must have property (b) of (2.3)(iii) i.e. its points must be in 1-1 correspondence with the pairs (V,L) (up to isomorphism). For the quotient $\alpha: \mathbb{A}^{n^2} \to \mathbb{A}^n$, this translates into the property that each fiber must be an orbit. But fibers are always closed, while from the closed orbit Lemma (1.7) orbits need not be closed (they are only locally closed).

Now $GL(n,k)$ does not act on $\mathbb{A}^{n^2}$ with closed orbits. For example, let F be an upper triangular $n \times n$ matrix and set

$$g_t = \begin{pmatrix} t^{r_1} & & & 0 \\ & t^{r_2} & & \\ & & \ddots & \\ 0 & & & t^{r_n} \end{pmatrix}$$

where $r_1 > \dots > r_n > 0$, all the r_i integers. Then as $t \to 0$, $g_t F g_t^{-1}$ approaches the semi-simple part of F. Thus unless F is already diagonal the orbit

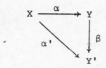

$O(F)$ is not closed.

Hence a coarse moduli space cannot exist for $\mathfrak{F}$.

Exercise. For the quotient $\alpha: \mathbb{A}^{n^2} \to \mathbb{A}^n$ show that each fiber contains a unique closed orbit consisting of semi-simple matrices, and a unique relatively open orbit consisting of matrices with a cyclic vector. These orbits coincide if all the eigenvalues are distinct.

For moduli space problems we must clearly strengthen the notion of quotient. As we have seen the key property is closedness of orbits. This leads us to the following definition:

Definition (3.6). Let G be an algebraic group acting on a variety X . A geometric quotient is a pair (Y,φ) , $\varphi: X \to Y$ morphism of varieties such that

(i) for every $y \in Y$, $\varphi^{-1}(y)$ is an orbit;

(ii) for each invariant open subset $U \subset X$, there exists an open subset $U' \subset Y$ such that $U = \varphi^{-1}(U')$;

(iii) for every open subset $U' \subset Y$, $\varphi^*: \mathcal{O}(U') \to \mathcal{O}(\varphi^{-1}(U'))$ defines an isomorphism of $\mathcal{O}(U')$ onto the ring of invariant functions $\mathcal{O}(\varphi^{-1}(U'))^G$ of $\varphi^{-1}(U')$.

Exercise. Show a geometric quotient is a quotient.

Remarks (3.7).

(i) Note that the geometric quotient has the property that its points parametrize the orbits and thus captures that property of what one should mean by "quotient space" geometrically. (No pun intended!)

(ii) Let $V_{n,m} = \{(F,G) \text{ completely reachable}\}$ as in (1.8). In Part IV we will show the Kalman quotient space $\mathcal{M}_{n,m} := V_{nm}/GL(n,k)$ is a geometric quotient. In (1.8)(ii) we have already indicated that $GL(n,k)$ acts on $V_{n,m}$ with closed orbits.

§4. Reductive Groups and Hilbert's 14th Problem

The notion of quotient of Section 3 is closely related to some topics of classical invariant theory as well as modern geometric invariant theory. We would like to make the connection precise in this section.

Definition (4.1). We say that an algebraic group G acts rationally on a vector space V of dimension n , if the action induces a morphism of algebraic groups $G \to GL(V)$.

We now make the temporary assumption that $\operatorname{char} k = 0$. A bit later we will discuss the corresponding theory in positive characteristics. The fundamental definition we will need for our invariant theory is:

Definition (4.2). A linear algebraic group G is linearly reductive if each rational action of G on any finite dimensional vector space V is completely reducible, i.e. if $W \subseteq V$ is an invariant subspace, then there is an invariant subspace $W' \subseteq V$ such that $V = W \oplus W'$.

Examples (4.3).

(i) $GL(n,k)$ is linearly reductive. For a proof see Fogarty [32], page 146.

(ii) Any semi-simple group is linearly reductive. See Fogarty [32], page 141.

(iii) A non-reductive group is for example the unipotent group discussed in (1.2) (ii).

We now make contact with the quotients of the previous section via the following beautiful theorem of Mumford [104], pages 27-30:

Theorem (4.4). Let X be an affine variety, G a linearly reductive group acting on X. Then an affine quotient (Y,φ) exists. Moreover G acts on X with closed orbits if and only if (Y,φ) is a geometric quotient.

Proof. We give the main idea of the proof in order to make contact with classical invariant theory and to motivate our discussion below. For details see [104].

Let $R = \mathcal{O}(X)$ be the coordinate ring of X. Then R is a finitely generated k-algebra. Let R^G be the ring of invariant functions. Then clearly R^G is also a k-algebra. It is the hypothesis that G is linearly reductive that allows one to conclude that R^G is finitely generated. This being so, R^G determines an affine variety Y and the inclusion $R^G \hookrightarrow R$ defines a morphism $\varphi: X \to Y$. It is then not too difficult to show that (Y,φ) is the required quotient.

Q.E.D.

Remark (4.5). Mumford's Theorem (4.4) is closely related to Hilbert's 14th problem. Specifically $GL(n,k)$ acts on k^n and hence on the symmetric algebra of k^n i.e. on $k[X_1,\ldots,X_n]$. Let $G \subset GL(n,k)$ be a linear algebraic group. Hilbert's 14th problem is to ask if $k[X_1,\ldots,X_n]^G$ is always a finitely generated k-algebra. The answer is no; a counter-example was given in 1958 by Nagata. See Nagata [109].

Mumford's theorem however gives an affirmative answer for G linearly reductive. This is why linearly reductive groups are especially important in invariant theory.

Now the above theory generalizes to a large extent to arbitrary algebraically closed fields k:

Definition (4.6). A linear algebraic group G (defined over $k = \bar{k}$ of arbitrary characteristic) is called reductive if its largest connected normal solvable subgroup (i.e. its radical) is a torus.

Remark (4.7). In characteristic 0 reductive groups are linearly reductive. For a proof see Fogarty [32], page 145. In characteristic $p > 0$ this is no longer true. Thus one needs a generalization of the notion of linear reductivity, this is given in the following definition (see Seshadri [131]) :

Definition (4.8). An algebraic group G is called geometrically reductive if given a finite dimensional vector space V on which G acts rationally, if V_0 is an invariant codimension 1 subspace, there exists an n for which the codimension 1 invariant subspace $V_0 \cdot \mathrm{Symm}^{n-1}V \subset \mathrm{Symm}^n V$ has an invariant 1-dimensional complement.

Remark (4.9). Mumford points out in the introduction to [104] that it is precisely the property of "geometric reductivity" which makes his geometric invariant theory and in particular the crucial theorem (4.4) go through. Of course if a group is linearly reductive, then it is geometrically reductive.

As mentioned before, in characteristic $p > 0$ reductive groups (in the sense of (4.6)) like $GL(n,k)$ need not be linearly reductive. Mumford conjectured however that reductivity implies geometric reductivity in every characteristic and thus that his theory is valid in every characteristic. Some partial results were proven (see Seshadri [130]) and finally in 1974 Haboush [50] proved the conjecture in full generality. It is probably the most important result in modern invariant theory.

The proof is beyond the scope of these notes but the interested reader can see Seshadri's nice description in the notes to the Arcata Conference in algebraic geometry [132].

§5. Richardson's Criterion

In this section we prove that the morphism $\alpha: \mathbb{A}^{n^2} \to \mathbb{A}^n$ defined by sending a matrix to its characteristic coefficients is a quotient. To do this we will use Richardson's criterion [117].

We first state without proof Richardson's lemma, a nice proof of which may be found in Kraft [89], page 113 :

Lemma (5.1). Let X , Y be irreducible affine varieties and suppose Y is normal (i.e. the coordinate ring of Y is integrally closed in its field of quotients). Let $\varphi: X \to Y$ be a surjective morphism which is an isomorphism on non-empty Zariski open subsets of X and Y . Then φ is everywhere an isomorphism.

Using (5.1) we can now prove Richardson's criterion ([117] and [89]).

Theorem (5.2). Let G be a linearly reductive algebraic group acting on an irreducible variety X . Let Y be a normal variety, and $\varphi: X \to Y$ a surjective morphism which is constant on the orbits. Then if there exists a dense subset $U \subset Y$ such that the fiber $\varphi^{-1}(y)$ contains exactly one closed orbit for $y \in U$, (Y,φ) is a quotient for X with respect to G .

Proof. By Mumford's Theorem (4.4) X has a quotient $(\tilde{Y},\tilde{\varphi})$ with respect to G . Thus by the universal property of quotients (3.3)(ii) there exists a unique morphism $\lambda: \tilde{Y} \to Y$ such that the diagram

$$\begin{array}{ccc} X & \xrightarrow{\varphi} & Y \\ & \searrow^{\tilde{\varphi}} & \uparrow \lambda \\ & & \tilde{Y} \end{array}$$

commutes.

It is clear that λ is surjective (from the surjectivity of φ). We claim that $\lambda^{-1}(y)$ consists of one point for all $y \in U$. To see this note by hypothesis that $\varphi^{-1}(y)$ contains a unique closed orbit and hence must be contained in a fiber of $\tilde{\varphi}$, which by the commutativity of the diagram proves the claim. Hence $\lambda: \tilde{Y} \to Y$ satisfies the hypotheses of Richardson's Lemma (5.1) from which we may conclude the proof.

Q.E.D.

Corollary (5.3). Let $\alpha: \mathbb{A}^{n^2} \to \mathbb{A}^n$ be defined by $\alpha(A) = (c_1(A),\ldots,c_n(A))$ where the $c_i(A)$ are the characteristic coefficients of A . Then $(\mathbb{A}^n,\alpha)$ is a $GL(n,k)$ quotient for $\mathbb{A}^{n^2}$.

Proof. We apply (5.2). Let $Q = \{A \in \mathbb{A}^{n^2} | A$ has distinct eigenvalues$\}$. Q is a dense open subset of $\mathbb{A}^{n^2}$ and for every $A \in Q$, $\alpha^{-1}(\alpha(A))$ consists of one orbit. Moreover $U: = \alpha(Q)$ is dense in $\mathbb{A}^n$. Hence by (5.2) we are done.

Q.E.D.

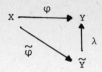

PART IV. GLOBAL MODULI OF LINEAR TIME-INVARIANT DYNAMICAL SYSTEMS

In Part II, Section 5 we noted that because of the redundancy in the state space description of a linear time-invariant dynamical system Σ of dimension n over a field k, in order to describe the input/output behavior (or equivalently to derive a coordinate-free description) one must look at the natural $GL(n,k)$-action on Σ induced by change of basis in the state space. Let us assume $k = \bar{k}$. Now the space of all such systems Σ of dimension n with space of input values $U \cong k^m$, and space of output values $Y \cong k^p$ may be identified with the space of all matrix triples $\{(F,G,H) | F \; n\times n \quad G \; n\times m \quad H \; p\times n\}$ which in turn is isomorphic to $k^{n^2+nm+np}$. $GL(n,k)$ acts on $k^{n^2+nm+np}$ by $(F,G,H) \longmapsto (gFg^{-1}, gG, Hg^{-1})$ for $g \in GL(n,k)$. We temporarily just consider the control part of the dynamical systems i.e. we consider the space of pairs $\{(F,G)\} \cong k^{n^2+nm}$. One wants to study then the geometric properties of the orbit space $k^{n^2+nm}/GL(n,k)$. The problem is that such orbit spaces are usually very ugly, in particular may be highly singular and even non-separated.

Now part of the power of Mumford's geometric invariant theory [104] is that it shows how one may identify a Zariski open subset of k^{n^2+nm} invariant under the action of $GL(n,k)$ on which a geometric quotient space exists with nice algebraic properties (this will all be made precise below). Moreover this subspace of k^{n^2+nm} naturally identified by geometric invariant theory is precisely the subspace of completely reachable pairs, a subspace which is naturally identified by system theory. It is this fact that makes Kalman's construction [79] of the quotient space of completely reachable pairs work and even implies the moduli properties of this space.

In short, the algebro-geometric properties of quotient spaces of completely reachable pairs and full dynamical systems and especially their relation to algebraic system theory will be our topic in this part of the lectures.

§1. Complete Reachability and Pre-Stability

In this section we consider the problem of giving a natural invariant-theoretic interpretation to the system-theoretic notion of "complete reachability". For other treatments of this subject see the very nice papers, Byrnes [17], Byrnes-Gauger [20], and Byrnes-Hurt [21]. The treatment we follow here is from Tannenbaum [148].

The first problem we will consider is the computation of the dimension of the stabilizer of a pair of matrices (F,G) over an <u>arbitrary</u> field k, $F \; n\times n$, $G \; n\times m$, relative to the action of $GL(n,k)$ given by $(F,G) \longmapsto (gFg^{-1}, gG)$ for

$g \in GL(n,k)$. We will translate this into a module-theoretic problem which leads us to our first proposition:

Proposition (1.1). *Let R be a commutative ring with identity, V an R-module, and $V' \subseteq V$ an R-submodule. Denote by $\mathcal{A}$ the group of R-automorphisms of V which are the identity on V'. Then there exists a natural exact sequence of groups*

$$0 \to \mathrm{Hom}_R(V/V',V') \xrightarrow{u} \mathcal{A} \xrightarrow{v} \mathrm{Aut}_R(V/V')$$

where $\mathrm{Hom}_R(V/V',V')$ is given its natural structure as an additive group.

Proof. First we let $v\colon \mathcal{A} \to \mathrm{Aut}_R(V/V')$ be the natural homomorphism from $\mathcal{A}$ to $\mathrm{Aut}_R(V/V')$. Then

$$\ker v = \{g \in \mathcal{A} \mid gx-x \in V' \quad \text{for every} \quad x \in V\}.$$

We must show that $\ker v \cong \mathrm{Hom}_R(V/V',V')$ with $\mathrm{Hom}_R(V/V',V')$ given its natural structure of additive group. Define $\varphi\colon \ker v \to \mathrm{Hom}_R(V/V',V')$ by letting $\varphi(g)(\bar{x}) = gx-x$ for $g \in \ker v$, $\bar{x} \in V/V'$, and $x \in V$ a representative of $\bar{x}$. Then it is easy to check that φ is well-defined and $\varphi(gg') = \varphi(g)+\varphi(g')$.

To show φ is an isomorphism we define an inverse. First note we have an exact sequence of R-modules

$$0 \to V' \xrightarrow{i} V \xrightarrow{\pi} V/V' \to 0.$$

Define a homomorphism $\psi\colon \mathrm{Hom}_R(V/V',V') \to \mathcal{A}$ by $\psi(\alpha) = \mathrm{id} + i\circ\alpha\circ\pi$ where $\alpha \in \mathrm{Hom}_R(V/V',V')$, and id is the identity map on V. Then it is easy to check ψ is a homomorphism and moreover the inverse of φ.

Q.E.D.

Remarks (1.2).

(i) Let (F,G) be a pair of matrices over k with F $n \times n$, G $n \times m$ $(n \geq m)$. Let $V := k^n$ regarded as a $k[X]$-module via F, i.e. $X\cdot v = F(v)$ for $v \in V$. Let V' be the $k[X]$-submodule of V generated by the columns of G. Then clearly the stabilizer subgroup of $GL(n,k)$ relative to the action $(F,G) \longmapsto (gFg^{-1},gG)$, $\mathrm{stab}((F,G))$, is isomorphic to the group of $k[X]$-automorphisms of V which are the identity on V', i.e. the group $\mathcal{A}$ of (1.1). In particular $\dim \mathcal{A} = \dim \mathrm{stab}(F,G)$.

(ii) From the exact sequence of (1.1), we have immediately that if (F,G) is completely reachable (i.e. $V' = V$), then $\mathrm{stab}((F,G)) = \{\text{identity}\}$. Unfortunately over an arbitrary field k, the converse is false. For example over $k = \mathbb{Z}/2\mathbb{Z}$ consider

$$F = \begin{pmatrix} 0 & 0 \\ 0 & 1 \end{pmatrix}, \quad G = \begin{pmatrix} 0 \\ 1 \end{pmatrix}$$

Then $\mathrm{stab}((F,G)) = \{\text{identity}\}$, but (F,G) is not completely reachable.

In terms of the exact sequence of (1.1) the point is that we may have $V' \subsetneqq V$ but $\mathrm{Hom}_R(V/V',V') = 0$ and $\mathrm{Aut}_R(V/V') = 0$. This leads us to :

Proposition (1.3). Let R be a principal ideal domain, V a finitely generated torsion R-module, $V' \subsetneq V$ a non-zero R-subnodule. Then if V/V' is not isomorphic to $\mathbb{Z}/2\mathbb{Z}$, $\mathcal{A} \neq \{\text{identity}\}$.

Proof. We first set some notation. Let M be an arbitrary finitely generated torsion R-module. For $p \in R$ a non-zero prime set $M(p) = \{m \in M \mid p^s m = 0$ for some $s > 0\}$. Then it is well-known that $M = \bigoplus_{\substack{p \text{ prime} \\ p \neq 0}} M(p)$ and moreover $M(p) \cong R/p^{\nu_1} \oplus \dots \oplus R/p^{\nu_s}$ for $1 \leq \nu_1 \leq \dots \leq \nu_s$.

Finally note that if $p,q \in R$ are non-zero primes $(p \neq q)$ then $(p)+(q) = R$ and therefore the Chinese remainder theorem applies. In particular, $(R/p^{\mu})_q = 0$ for $p \neq q$ (where for any R-module M and any prime p, M_p denotes the localization of M at the prime ideal (p)). We divide the proof of (1.3) into two cases.

Case (1). For every p such that $V'_p \neq 0$, $V'_p = V_p$. Then from our above discussion this implies that $V \cong V' \oplus V/V'$ as R-modules and $\mathrm{Hom}_R(V/V',V') = 0$. Hence from (1.1) $\mathcal{A} \cong \mathrm{Aut}_R(V/V')$. But then it is an easy exercise to show that if V/V' is not isomorphic to $\mathbb{Z}/2\mathbb{Z}$, then $\mathrm{Aut}_R(V/V') \neq \{\text{identity}\}$. (Use the decompositions

$$V/V' \cong \bigoplus_{\substack{p \text{ prime} \\ p \neq 0}} (V/V')(p)$$

and $(V/V')(p) \cong R/p^{\mu_1} \oplus \dots \oplus R/p^{\mu_t}$.)

Case (2). There exists $p \in R$ a non-zero prime such that $0 \neq V'_p \subsetneq V_p$. Then there exists p such that $V'(p) \neq 0$ and $(V/V')(p) \neq 0$. From (1.1) it is enough to show that $\mathrm{Hom}(V/V',V') \neq 0$, so we are reduced to showing that $\mathrm{Hom}((V/V')(p),V'(p)) \neq 0$. But

$$(V/V')(p) \cong R/p^{\mu_1} \oplus \dots \oplus R/p^{\mu_t}$$

$$V'(p) \cong R/p^{\nu_1} \oplus \dots \oplus R/p^{\nu_s}.$$

If $\mu_1 > \nu_1$ we have the natural R-homomorphism

$$R/p^{\mu_1} \to R/p^{\nu_1}.$$

If $\mu_1 \leq \nu_1$ we have the homomorphism

$$R/p^{\mu_1} \xrightarrow{p^{\nu_1 - \mu_1}} R/p^{\nu_1}$$

defined by multiplication by $p^{\nu_1-\mu_1}$.

Q.E.D.

Corollary (1.4). If $k \neq \mathbb{Z}/2\mathbb{Z}$, then (F,G) is completely reachable if and only if $\mathrm{stab}(F,G) = \{\text{identity}\}$.

Proof. Immediate from (1.2) and (1.3) above.

Q.E.D.

We can now prove a result over an arbitrary closed field k first given by

Byrnes and Hurt in [21] for the case char $k = 0$:

Corollary (1.5). *Let* $k = \bar{k}$ *be an arbitrary algebraically closed field. Let* (F,G) *be a pair of matrices over* k , F $n \times n$, G $n \times m$. *Then* (F,G) *is completely reachable if and only if the stabilizer of* (F,G) *in* $GL(n,k)$ *is zero dimensional.*

Proof. From the above we are reduced to showing that if $\dim \mathrm{stab}((F,G)) = 0$, then $\mathrm{stab}((F,G)) = \{\text{identity}\}$. Let $M_{n,n}(k)$ be the set of $n \times n$ matrices over k . Clearly $L := \{A \in M_{n,n}(k) \mid AF = FA,\ AG = G\}$ is a linear subspace of $M_{n,n}(k)$, and $\mathrm{stab}((F,G)) = L \cap GL(n,k)$. But $\mathrm{stab}((F,G))$ is a closed algebraic subgroup of $GL(n,k)$, and being zero dimensional (by hypothesis), it must consist of only finitely many points. But L is irreducible and $L \cap GL(n,k)$ is dense in L , and so $\mathrm{stab}((F,G)) = L \cap GL(n,k)$ is irreducible. Thus $\mathrm{stab}((F,G))$ consists of a single point which must be the identity.

Q.E.D.

Corollary (1.6). *Let* $V_{n,m} := \{(F,G) \in M_{n,n}(k) \times M_{n,m}(k) \mid (F,G)$ *is completely reachable*$\}$. *Then* $GL(n,k)$ *acts on* $V_{n,m}$ *with closed orbits.*

Proof. Immediate from the closed orbit Lemma (1.7) of Part III and (1.5) above.

Q.E.D.

Remark (1.7). The problem of computing $\dim \mathcal{A} = \dim \mathrm{stab}(F,G)$ in general is still not solved. If one assumes however that $\mathcal{A}^{\circ}$ is solvable ($\mathcal{A}^{\circ}$ = connected component of the identity in $\mathcal{A}$) then using the Borel fixed point theorem [10] one can show that there exists a "flag" of $k[X]$-modules $V = V_n \supseteq V_{n-1} \supseteq \cdots \supseteq V_{n-r} = V'$ with dimension of $V_i = i$ $(i = n-r,\dots,n)$ and such that V_i/V_{i-1} is a simple $k[X]$-module with $\mathcal{A}^{\circ}V_i = V_i$ for each i . This then gives an inductive procedure for computing $\dim \mathcal{A}$. See also Part V, (3.3)(ii).

We now make contact with the geometric invariant theory of Part III of these notes in order to give an invariant-theoretic interpretation to the notion of "complete reachability". We first need the following technical lemma:

Lemma (1.8). Let X be a k-variety $(k = \bar{k})$, G a reductive group acting on X , and $U \subset X$ a connected invariant affine subset. Then if G acts on U with closed orbits, all the stabilizers of all the closed points of U have the same dimension.

Proof. First from Mumford's theorem (4.4) of Part III, a geometric quotient (Y,π) of U exists relative to G . By definition, the fibers of $\pi: U \to Y$ over the closed points of Y are precisely the orbits of the points of U . Set for each $x \in U$, $s(x) := \dim \mathrm{stab}(x)$, $d(x) := \dim O(x) = \dim \pi^{-1}(\pi(x))$ where $O(x)$ denotes the orbit of x in U . Then from [52] $s(x)$ and $d(x)$ are upper-semicontinuous functions on U . But since $G/\mathrm{stab}(x) \cong O(x)$, $\dim G = s(x)+d(x)$ for all $x \in U$, and thus by the connectivity of U , $s(x)$ and $d(x)$ are constant.

Q.E.D.

Remark (1.9). Let $V_{n,m}$ be the Zariski open subset of completely reachable pairs as in (1.6) above. Define an equivariant morphism $R: M_{n,n}(k) \times M_{n,m}(k) \to M_{n,mn}(k)$ by $R(F,G) = (G\ FG\ \ldots\ F^nG)$ (see Hazewinkel-Kalman [58] for details). Number the columns of $R(F,G)$ lexicographically as $01\ 02\ \ldots\ 0m\ 11\ \ldots\ 1m\ \ldots\ n1\ \ldots\ nm$, and define a nice selection to be a subset I of this set of indices of size n such that if $(i,j) \in I$, then $(i',j) \in I$ for all $i' \leq i$. Then a fundamental observation of Kalman [79] is that (F,G) is completely reachable if and only if there exists a nice selection I such that $\det R(F,G)_I \neq 0$.

Accordingly we define invariant open affine subsets of $M_{n,n}(k) \times M_{n,m}(k)$ for each nice selection I by

$$U_I := \{(F,G) \mid \det R(F,G)_I \neq 0\} .$$

Clearly the $\{U_I\}_{I \text{ nice}}$ form an open affine cover of $V_{n,m}$.

Finally we remark that from (1.6) it is clear that $GL(n,k)$ acts on U_I with closed orbits and thus from a theorem of Luna [94] $U_I \cong GL(k,n) \times k^{nm}$ and this isomorphism is equivariant. Hazewinkel-Kalman [58] give a direct construction of this isomorphism.

Before stating the main result of this section we need to make the following key definition (of course due to Mumford [104], page 36):

Definition (1.10). Let X be a k-variety, G a reductive group acting on X. Then a point $x \in X$ is pre-stable if there exists an invariant open affine subset $U \subseteq X$ such that $x \in U$ and G acts on U with closed orbits.

We can now state the main result of this section:

Theorem (1.11). Let $V_{n,m}$ be as above. Then $V_{n,m}$ is precisely the set of pre-stable points of $M_{n,n}(k) \times M_{n,m}(k) \cong k^{n^2+nm}$ relative to the action of $GL(n,k)$.

Proof. Let $P \subseteq k^{n^2+nm}$ be the Zariski open subset of pre-stable points. We first show $V_{n,m} \subseteq P$. But from (1.9) for every $(F,G) \in V_{n,m}$, there exists an invariant open affine subset U_I containing (F,G) on which $GL(n,k)$ acts with closed orbits, i.e. (F,G) is pre-stable.

Conversely let $(F,G) \in P$, and let U be an open invariant affine subset of k^{n^2+nm} containing (F,G) on which $GL(n,k)$ acts with closed orbits. Now $V_{n,m}$ is dense in k^{n^2+nm} so there exists $(F',G') \in V_{n,m} \cap U$. But from (1.8) the stabilizers of all the points in U have the same dimension so in particular $\dim \operatorname{stab}((F,G)) = \dim \operatorname{stab}((F',G')) = 0$ (since (F',G') is completely reachable). Hence by (1.5), $(F,G) \in V_{n,m}$.

Q.E.D.

Corollary (1.12). A geometric quotient $(\mathcal{M}_{n,m}, \pi)$ exists for $V_{n,m}$ relative to $GL(n,k)$.

Proof. Given (1.11), this follows immediately from Proposition (1.9) of Mumford [104], page 37.

Q.E.D.

Remarks (1.13).

(i) The specific structure of the geometric quotient space $\mathcal{M}_{n,m}$ will be the main topic of the remaining sections of Part IV. As we will see $\mathcal{M}_{n,m}$ is a moduli space generalizing the results of Part III, Section 2.

(ii) The action of $GL(n,k)$ on k^{n^2+nm} (considered to be $M_{n,n}(k) \times M_{n,m}(k)$) comes about by change of basis in the state space k^n. We may also consider change of basis in the input space k^m. This induces an action of $GL(m,k)$ on k^{n^2+nm} given by $(F,G) \longmapsto (F,Gh)$ for $h \in GL(m,k)$. Hence we get a natural action of $GL(n,k) \times GL(m,k)$ on k^{n^2+nm} by $(F,G) \longmapsto (gFg^{-1},gGh)$ for $g \in GL(n,k)$ and $h \in GL(m,k$. We leave it as an exercise for the reader to show that the set of pre-stable points of k^{n^2+nm} with respect to the action of Q is

$$\tilde{V}_{n,m} := \{(F,G) \mid (F,G) \text{ is completely reachable and } G \text{ is of maximal rank}\}.$$

§2. Construction of the Quotient Space of Completely Reachable Pairs

In this section we give an explicit construction of the quotient $\mathcal{M}_{n,m}$ of Section 1. We remark that many of the key ideas of this construction are already in the Kalman paper [79]. Other treatments of this topic can be found in Hazewinkel [54], Hazewinkel-Kalman [58], Byrnes-Hurt [21]. We prefer to use the geometric invariant theoretic ideas of Mumford [104] here in a rather straight-forward way. See specifically pages 38-39.

Construction of $\mathcal{M}_{n,m}$ (2.1). We use the notation of (1.9). Accordingly for I a nice selection let $U_I = \{(F,G) \mid \det R(F,G)_I \neq 0\}$. As noted before $GL(n,k)$ acts on the affine space U_I with closed orbits. Applying Mumford's theorem (Part III, (4.4)), there exists a geometric quotient (V_I,π_I) of U_I and V_I is affine.

Next given I,J any two nice selections, define an invariant function on U_I by

$$\sigma_{IJ}(F,G) = \det R(F,G)_J / \det R(F,G)_I .$$

Then σ_{IJ} being invariant descends to a function on the quotient V_I. Let

$$V_{IJ} = V_I - \{(F,G) \mid \sigma_{IJ}(F,G) = 0\} .$$

Clearly $\pi_I^{-1}(V_{IJ}) = U_I \cap U_J = \pi_J^{-1}(V_{JI})$. Since V_I (respectively V_J) is a quotient of U_I (respectively U_J), it follows that both V_{IJ} and V_{JI} are quotients of $U_I \cap U_J$ and thus by uniqueness we have a canonical isomorphism $\varphi_{IJ}: V_{IJ} \xrightarrow{\sim} V_{JI}$ such that the diagram

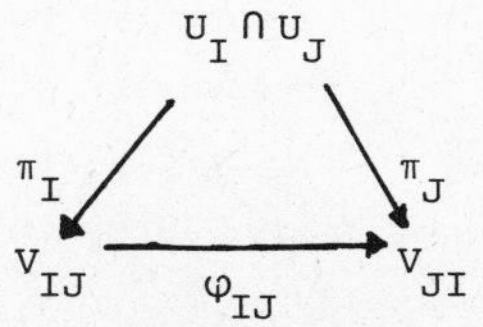

commutes. Thus the $\{V_I\}_{I \text{ nice}}$ patch together to form a variety $\mathcal{M}_{n,m}$.

Moreover $\{\sigma_{IJ}\}$ forms a Czech 1-cocycle and hence its image in $H^1(\mathcal{M}_{n,m}, \mathcal{O}^*_{\mathcal{M}_{n,m}})$ determines a line bundle L. (For a discussion of this cohomological treatment of line bundles see e.g. Mumford [105], pages 28-29.) Following [104], page 39, one can show that L is ample i.e. that for $N \gg 0$, the sections of L^N determine an embedding of $\mathcal{M}_{n,m}$ into some projective space. This means that $\mathcal{M}_{n,m}$ is quasi-projective. We summarize our discussion with the following theorem:

Theorem (2.2). $\mathcal{M}_{n,m}$ is a smooth irreducible quasiprojective variety of dimension nm such that $\pi: V_{n,m} \to \mathcal{M}_{n,m}$ is a geometric quotient. Moreover $V_{n,m}$ is a principal algebraic $GL(n,k)$-bundle over $\mathcal{M}_{n,m}$.

Proof. The first part is immediate from (1.9), (1.12) and (2.1). The last assertion is immediate from Mumford [104], page 16.

Q.E.D.

Remarks (2.3).

(i) There is a related concept to that of pre-stability in geometric invariant theory, namely that of stability (see [104], page 36). Then while there are no $GL(n,k)$-stable points of k^{n^2+nm} (see [21] or [148]), Byrnes-Hurt [21], pages 108-110 prove that $V_{n,m}$ is the set of $SL(n,k)$-stable points of k^{n^2+nm}.

(ii) Theorem (2.2) is deceptive in the following sense: While we know that $\mathcal{M}_{n,m}$ is quasi-projective, the possibility still remains that $\mathcal{M}_{n,m}$ may be quasi-affine (i.e. be embedded in some affine variety as an open subset). In point of fact by the existence of the sharp control canonical form (see the proof of (2.9) in Part III) we clearly have $\mathcal{M}_{n,1} \cong k^n$. In Section 4 we will show that for $m > 1$, $\mathcal{M}_{n,m}$ is never quasi-affine and for any m, $\mathcal{M}_{n,m}$ is never projective.

§3. Moduli of Linear Time-Invariant Dynamical Systems

Recalling our discussion in Section 2 of Part III (see especially (2.8) and (2.9)) it should be crystal clear that the Kalman quotient must have moduli space properties. Indeed looking at (2.8) of Part III it should even be clear what functor $\mathcal{M}_{n,m}$ must represent. This generalization of Mumford's results given in Part III was carried out by Hazewinkel (see [53] and [54]) and leads us to the following definition:

Definition (3.1). Let $\mathcal{F}$: Var(k) $\to$ Sets be the contravariant functor from the category of varieties over k to the category of sets given for $S \in \text{Var}(k)$ by

$\mathcal{F}(S)$: = {isomorphism classes of completely reachable families of pairs with m inputs and state space dimension n over S, i.e. isomorphism classes of $(m+2)$-tuples $(\mathcal{B}, \tilde{F}, g_1, \ldots, g_m)$ where $\mathcal{B}$ is a rank n vector bundle over S, $\tilde{F}$ is an endomorphism of $\mathcal{D}$, and $g_1,\dots,g_m$ global sections such that for each point $s \in S$, $\tilde{F}^i(s)g_j(s)$ generate the fiber $\mathcal{D}(s)$ for $0 \leq i \leq n-1$, $1 \leq j \leq m\}$.

Remarks (3.2).

(i) Note that one has a natural morphism of functors from $\mathfrak{F} \to h_{\mathcal{M}_{n,m}}$. Indeed for $S \in \mathrm{Var}(k)$, given an element $(\mathcal{D},\tilde{F},g_1,\dots,g_m)$ of $\mathfrak{F}(S)$, for every $s \in S$ after choosing a basis in $\mathcal{D}(s)$ we get a completely reachable pair $(F(s),G(s))$. This pair is unique modulo a choice of basis and thus determines a point of $\mathcal{M}_{n,m}$. Hence $(\mathcal{D},\tilde{F},g_1,\dots,g_m)$ determines a morphism $S \to \mathcal{M}_{n,m}$ and thus in turn we have our morphism of functors $\mathfrak{F} \to h_{\mathcal{M}_{n,m}}$.

(ii) In [54] Hazewinkel shows that $\mathfrak{F}(S) \cong \mathrm{Hom}(S,\mathcal{M}_{n,m})$, i.e. $\mathcal{M}_{n,m}$ represents the functor $\mathfrak{F}$ and thus is a fine moduli space. Hazewinkel proves this by constructing a universal completely reachable family of pairs over $\mathcal{M}_{n,m}$. More specifically, in order to show that $\mathfrak{F}(S) \cong \mathrm{Hom}(S,\mathcal{M}_{n,m})$ it suffices to construct a family Σ_u of completely reachable pairs over $\mathcal{M}_{n,m}$, such that for any completely reachable family Σ over a variety S, there exists a unique morphism $F\colon S \to \mathcal{M}_{n,m}$ such that $\Sigma \cong f^* \Sigma_u$. (The definition of the pull-back f^* should be clear.) Indeed this universal family Σ_u corresponds to the identity morphism $\mathcal{M}_{n,m} \to \mathcal{M}_{n,m}$. (Of course to show for any contravariant functor $\mathfrak{F}\colon \mathrm{Var}(k) \to \mathrm{Sets}$ that a certain variety M represents it, one need only construct the universal object as above which will correspond to the identity morphism $M \to M$.) Actually Hazewinkel constructs universal families over the V_I (notation as in (2.1)) and then by explicit computation shows that these patch together to form a global universal family over $\mathcal{M}_{n,m}$. Now the proof we give below combines Hazewinkel's local constructions with the techniques of Kleiman [86] showing the universality of the Grassmannian. This has the advantage of showing that the local pieces patch together automatically (one does not need to explicitly write down patching data) and moreover given the close relationship between $\mathcal{M}_{n,m}$ and the Grassmannian $\mathrm{Gr}(n,(n+1)m)$ (see Section 5 below), this seems the most natural way to proceed.

Theorem (3.3). Notation as above. Then $\mathcal{M}_{n,m}$ is a fine moduli space, i.e. $\mathfrak{F}(S) \cong \mathrm{Hom}(S,\mathcal{M}_{n,m})$ for all $S \in \mathrm{Var}(k)$.

Proof. First from the definition of $\mathfrak{F}$, it is clear that for given $S \in \mathrm{Var}(k)$, for $U \subseteq S$ open, the presheaf of sets $U \to \mathfrak{F}(U)$ is a sheaf. Now given a matrix pair $(F,G) \in M_{n,n}(k) \times M_{n,m}(k)$ if $R(F,G) = (G\ FG \dots F^{n-1}G)$ let I be a nice selection of the columns. For each such I define a subfunctor $\mathfrak{F}_I$ of $\mathfrak{F}$ by the condition for $S \in \mathrm{Var}(k)$,

$\mathcal{F}_I(S)$ = {isomorphism classes of completely reachable families of pairs over S with m inputs and state space dimension n, $(\mathcal{D},\tilde{F},g_1,\ldots,g_m)$, such that for each point $s \in S$, $\tilde{F}(s): \mathcal{D}(s) \to \mathcal{D}(s)$. $g_1(s),\ldots,g_m(s) \in \mathcal{D}(s)$ are such that for some (and hence every) choice of basis in $\mathcal{D}(s)$, the corresponding completely reachable pair $(F(s),G(s))$ has the property that $\det(R(f(s),G(s))_I) \neq 0$} .

Note that the $\mathcal{F}_I$ are open subfunctors of $\mathcal{F}$ in the sense that if $(\mathcal{D},\tilde{F},g_1,\ldots,g_m) \in \mathcal{F}(S)$ (up to isomorphism) and $s \in S$ is a point with the corresponding completely reachable pair (after choosing a basis in $\mathcal{D}(s)$) having the property that $\det R(F(s),G(s))_I \neq 0$, then there exists an open neighborhood U of s such that $(\mathcal{D},\tilde{F},g_1,\ldots,g_m)|U$ determines an element of $\mathcal{F}_I(U)$.

Next it is almost trivial to prove that the functors $\mathcal{F}_I$ are representable and in point of fact if (V_I,π_I) is the geometric quotient of U_I (notation as in (1.9) and (2.1)), then $\mathcal{F}_I \cong h_{V_I}$. We can use the same argument as that given in [54]. Indeed first note that since $U_I \cong GL(n,k) \times k^{nm}$ equivariantly (see (1.9)), the quotient $\pi_I: U_I \to V_I$ admits a section σ_I (in the language of [54] on U_I we have an "algebraic canonical form"). Now as remarked in (3.2)(ii), it suffices to construct a universal completely reachable family over V_I which represents an element of $\mathcal{F}_I(V_I)$. Since $V_I \cong k^{nm}$, by the Quillen-Suslin theorem [116], the only choice we have for a rank n vector bundle over V_I is the trivial bundle $V_I \times k^n$. Setting $(F(x),G(x)) := \sigma_I(x)$ we define an endomorphism $F_I: V_I \times k^n \to V_I \times k^n$ by $\tilde{F}_I(x,y) := (x,F(x)y)$, and sections $g_{Ii}: V_I \to V_I \times k^n$ by $x \longmapsto (x$, i-th column of $G(x))$ for $i = 0,\ldots,m$. It is immediate that $(V_I \times k^n, \tilde{F}_I, g_{I1},\ldots,g_{Im})$ is a universal family.

Now note that the $\{\mathcal{F}_I\}_{I \text{ nice}}$ forms an open covering of $\mathcal{F}$ in the sense that for every $S \in Var(k)$ if $(\mathcal{D},F,g_1,\ldots,g_m) \in \mathcal{F}(S)$ (up to isomorphism) and $s \in S$ is a point, then there exists a nice selection I and an open neighborhood U of s such that $(\mathcal{D},F,g_1,\ldots,g_m)|U$ determines an element of $\mathcal{F}_I(U)$.

We can now finish the argument as in Kleiman [86], page 283. Indeed letting $\mathcal{F}_{IJ} = \mathcal{F}_I \times_{\mathcal{F}} \mathcal{F}_J$ we see that $\mathcal{F}_{IJ} \to \mathcal{F}_I$ is an open immersion of varieties (where we identify the functor $\mathcal{F}_I$ with the variety V_I which represents it). From the fact that $\mathcal{F}$ is covered by the open representable subfunctors $\mathcal{F}_I$, the $\mathcal{F}_I$ patch along the $\mathcal{F}_{IJ}$ to form a variety, and since $\mathcal{F}$ is a "Zariksi sheaf" (i.e. $U \to \mathcal{F}(U)$ is a sheaf for $U \subseteq S$ open), this variety must represent $\mathcal{F}$. Finally since $\mathcal{M}_{n,m}$ and $\mathcal{F}$ are patched together in the same way, it is clear that it is $\mathcal{M}_{n,m}$ which represents $\mathcal{F}$.

Q.E.D.

<u>Remark (3.4).</u> Hazewinkel has shown in [57] that $\mathcal{M}_{n,m}$ may be defined over any commutative ring with identity, e.g. $\mathbb{Z}$, by local patching arguments. Again the preceding techniques of Kleiman will lead to a proof of this fact. (Kleiman

[86] shows that the Grassmannian is defined over any commutative ring with identity.)

§4. The Geometric Structure of the Moduli Space

In this section we would like to show that for $m > 1$, $\mathcal{M}_{n,m}$ is never quasi-affine. This is closely related to Hazewinkel's result on the non-existence of global continuous canonical forms and as we shall see Hazewinkel's result is a simple corollary of the first result. We identify throughout this section k^{n^2+nm} with $M_{n,n}(k) \times M_{n,m}(k)$. Finally we would like to thank Hanspeter Kraft for explaining to us the technique used below of first taking an $SL(n,k)$ quotient and then a k^* quotient of k^{n^2+nm} since it is this technique which pushes everything through.

Remark (4.1). First note in the standard way we can take $GL(n,k) = SL(n,k)\cdot k^*$. Set $X := k^{n^2+nm}$, and let (Y,π) be the quotient of X by $SL(n,k)$ (which exists by Mumford's Theorem (4.4) of Part III). Then if X' denotes the subset of X fixed by k^*, it is clear $X' \cong k^{n^2}$. Now set $Y' := \pi(X')$. From (5.3) of Part III, $Y' \cong k^n$.

Clearly Y is a k^*-variety and if $O_{k^*}(y)$ denotes the orbit of $y \in Y$ under k^* then $\overline{O_{k^*}(y)} = O_{k^*}(y) \cup \{\overline{O_{k^*}(y)} \cap Y'\}$, and one may check that $\overline{O_{k^*}(y)} \cap Y'$ consists of a single point. In this circumstance we have the following standard lemma:

Lemma (4.2). Let $\tilde{Y} := Y - Y'$. Then a geometric quotient $\tilde{Y}/k^*$ exists for $\tilde{Y}$ relative to k^*, and moreover there exists a natural projective morphism $\varphi: \tilde{Y}/k^* \to Y'$.

Proof. The existence of the geometric quotient is obvious from our above remarks. Next recall that in general for two varieties A_1 and A_2 a morphism $\lambda: A_1 \to A_2$ is projective if λ may be factored by a closed immersion of A_1 into $\mathbb{P}^N \times A_2$ (for some $N \geq 0$) followed by the natural projection onto A_2.

In our case the statement about the existence of a projective morphism $\varphi: \tilde{Y}/k^* \to Y'$ is a straightforward generalization of the map $k^r - (0)/k^* \to (0)$ be being projective (of course here $k^r - (0)/k^* = \mathbb{P}^{r-1}$!). Technically (for the relevant definitions see Hartshorne [52]), the fact that Y is a k^* variety means that the coordinate ring $\mathcal{O}(Y) = \bigoplus_{i \geq 0} \mathcal{O}(Y)_i$ has a graded structure. Then the morphism φ is nothing but $\text{Proj}\ \mathcal{O}(Y) \to \text{Spec}\ \mathcal{O}(Y)_o$.

Q.E.D.

Remark (4.3). Now put $\tilde{X} = X - \pi^{-1}(\pi(X'))$. (Notation as in (4.1). Then we have a commutative diagram

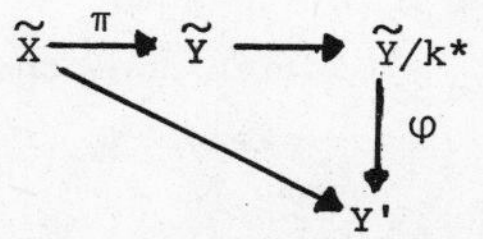

with φ projective. It is easy to check from our above remarks that $\tilde{Y}/k^*$ is precisely $\mathcal{M}_{n,m}$. See also [148].

We summarize our above discussion with the following theorem:

Theorem (4.4). *There exists a projective morphism* $\varphi: \mathcal{M}_{n,m} \to k^n$ *which makes the following diagram commutative:*

$$\begin{array}{ccccc} V_{n,m} & \hookrightarrow & k^{n^2+nm} & \longrightarrow & k^{n^2} \\ \downarrow \pi & & & & \downarrow \alpha \\ \mathcal{M}_{n,m} & & \longrightarrow & & k^n \end{array}$$

where α *is the quotient map of* (5.3) *of Part II.*

Proof. Immediate from the above discussion.

Q.E.D.

Corollary (4.5). *For* $m > 1$, $\mathcal{M}_{n,m}$ *is not quasi-affine.*

Proof. Indeed φ has projective varieties of positive dimension as fibers (see also Section 6), so $\mathcal{M}_{n,m}$ cannot lie in any affine space.

Q.E.D.

Remark (4.6). From (4.5) we will now derive trivially two important corollaries. The first will show that there exist no global algebraic canonical forms for $m > 1$. (For this terminology see Hazewinkel [54]). This condition means that π has no sections (i.e. morphisms $\mathcal{M}_{n,m} \overset{\sigma}{\to} V_{n,m}$ such that $\pi \cdot \sigma$ = identity). For $m = 1$, of course $\mathcal{M}_{n,1} \cong k^n$ and $V_{n,1} \cong GL(n,k) \times k^n$ by the existence of the control canonical form (Part II, Section 4).

The second corollary concerns a question discussed in Byrnes-Hurt [21], pages 102-103 (see (3.3)). They show for $n = 1$, $n = 2$ that $\mathcal{M}_{n,m}$ is not projective and ask when $\mathcal{M}_{n,m}$ is projective. We will show that $\mathcal{M}_{n,m}$ is *never* projective.

Corollary (4.7). *For* $m > 1$, *there exist no global algebraic canonical forms.*

Proof. From (4.6) we must show that for $m > 1$, $\pi: V_{n,m} \to \mathcal{M}_{n,m}$ has no sections. Suppose $\sigma: \mathcal{M}_{n,m} \to V_{n,m}$ were a section. Then in particular σ would be injective so that $\mathcal{M}_{n,m}$ would admit an immersion into the quasi-affine space $V_{n,m}$. But this cannot be by (4.5).

Q.E.D.

Corollary (4.8). $\mathcal{M}_{n,m}$ *is never projective.*

Proof. Indeed $\varphi: \mathcal{M}_{n,m} \to k^n$ is a non-constant morphism. Composing φ with the projection of $k^n \to k$ we see $\mathcal{M}_{n,m}$ admits non-constant global regular functions. Thus by Part I, Section 3 of these notes, $\mathcal{M}_{n,m}$ cannot be projective.

Q.E.D.

§5. The Global Moduli of Completely Reachable Matrix Triples

The investigation of the global moduli of completely reachable finite dimensional linear dynamical systems Σ , or equivalently triples of matrices (F,G,H) where (F,G) is a completely reachable matrix pair with F $n\times n$, G $n\times m$, and H $p\times n$, has been carried out in such various works as Hazewinkel [53], [54], [57], Byrnes [17] and Byrnes-Hurt [21]. There is some nice geometry here and so we would like to discuss some of these results. We will work over the field $\mathbb{C}$ of complex numbers for convenience even though many of these constructions work an arbitrary field and even over a commutative ring with unity ([57]). We first make a few preliminary remarks:

$\mathcal{M}_{n,m}$ and the Grassmannian (5.1).

(i) It should be clear that the moduli space $\mathcal{M}_{n,m}$ is closely connected to the Grassmannian. In point of fact it is completely elementary to show that there exists an immersion $\bar{R}: \mathcal{M}_{n,m} \to Gr(n,(n+1)m)$ induced by the $GL(n,\mathbb{C})$-equivariant morphism $R: V_{n,m} \to M^{reg}_{n,(n+1)m}$ defined in (1.9). (Here $M^{reg}_{n,(n+1)m}$ denotes $n\times(n+1)m$ matrices of maximal rank n .) Indeed to see that R is injective just note that if $(G\ FG\ \dots\ F^nG) = (G'\ F'G'\ \dots\ F'^nG')$, then $G = G'$, $F^iG = F'^iG$ for $i = 1,\dots,n$, and so $F(G\ FG\ \dots\ F^{n-1}G) = F'(G\ FG\ \dots\ F^{n-1}G)$ which means $F = F'$ since $(G\ FG\ \dots\ F^{n-1}G)$ has maximal rank. (Note that $G \neq 0$ since $(F,G) \in V_{n,m}$, and thus by the Cayley-Hamilton theorem the fact that $R(F,G)$ has maximal rank implies that $(G\ FG\ \dots\ F^{n-1}G)$ has maximal rank.) Next a simple computation shows that the differential $dR_{(F,G)}$ is non-singular. Indeed it is enough to check this at points of the form $(A,0)$ and then $dR_{(F,G)}(A,0) = (0\ AFG\ \dots\ AF^{n-1}G)$. By the inverse function theorem then R is an immersion.

Next R being $GL(n,\mathbb{C})$-equivariant implies that $\bar{R}$ descends to the $GL(n,\mathbb{C})$-quotients to define an immersion $\bar{R}: \mathcal{M}_{n,m} \to Gr(n,(n+1)m)$. By the way, from Chow's theorem [126], since $Gr(n,(n+1)m)$ is projective, this gives another proof without geometric invariant theory that $\mathcal{M}_{n,m}$ is quasi-projective. Note that we have a commutative diagram

$$\begin{array}{ccc} V_{n,m} & \overset{R}{\hookrightarrow} & M^{reg}_{n,(n+1)m} \\ \downarrow{\scriptstyle \pi} & & \downarrow{\scriptstyle P} \\ \mathcal{M}_{n,m} & \overset{\bar{R}}{\hookrightarrow} & Gr(n,(n+1)m) \end{array}$$

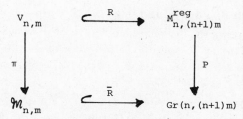

Moreover we will soon see below that $(M^{reg}_{n,(n+1)m},p)$ is a principal $GL(n,\mathbb{C})$ fiber bundle. In (2.2) we have shown already that $(V_{n,m},\pi)$ is an algebraic principal $GL(n,\mathbb{C})$ bundle.

(ii) Just as we were able to read off the geometric and moduli properties of $\mathfrak{M}_{n,m}$ from Mumford [104], [108], it will turn out that we will be able to read off the geometric and moduli properties of completely reachable triples again from these references plus from the descriptions given of the universal bundle over the Grassmannian in Griffiths-Adams [44], pages 100-103. Since this description is so crucial we give it now in some detail.

We use the notation of Part I, Section 6. So again for $r \leq s$, we have the standard morphism $p: M^{reg}_{r,s} \to Gr(r,s)$. Then for each multi-index $I = \{1 \leq i_1 < \ldots < i_r \leq s\}$ we have the Zariski open subset

$$U_I = \{\Lambda \in Gr(r,s) \mid \Lambda \text{ admits a matrix representation } A \text{ such that the I-th } r \times r \text{ minor of } A \text{ is non-singular}\}.$$

Set $V_I := p^{-1}(U_I)$. Then one defines a morphism $\varphi_I: V_I \to GL(r,\mathbb{C})$ by sending a matrix A to its I-th $r \times r$ minor. Then clearly the functions $\varphi_{IJ}: V_I \cap V_J \to GL(r,\mathbb{C})$ defined by $\varphi_{IJ}(A) := \varphi_I(A)^{-1}\varphi_J(A)$ descend to $U_I \cap U_J$ and satisfy the cocycle condition, and thus define a vector bundle of rank r on $Gr(r,s)$ called the <u>universal bundle</u> U_r.

Now we consider once again explicitly the principal bundle structure on $p: M^{reg}_{r,s} \to Gr(r,s)$. Recall in Part I, Section 6, we constructed isomorphisms $\pi_I: U_I \to \mathbb{C}^{r(s-r)}$. These isomorphisms can be lifted to give a commutative diagram

$$\begin{array}{ccc} V_I & \xrightarrow{\ \tilde{\pi}_I\ } & GL(r,\mathbb{C}) \times \mathbb{C}^{r(s-r)} \\ \downarrow & & \downarrow \\ U_I & \xrightarrow{\ \pi_I\ } & \mathbb{C}^{r(s-r)} \end{array}$$

and moreover it is immediate from the definitions that the transition maps $\tilde{\pi}_I \circ \tilde{\pi}_J^{-1}$ are defined by the φ_{IJ}^{-1} for each I,J. This description precisely means that the <u>dual</u> of the universal bundle U_r^*, is the $\mathbb{C}^r$-bundle associated to the principal $GL(r,\mathbb{C})$-bundle $M^{reg}_{r,s} \to Gr(r,s)$. (Recall that if E is a principal G-bundle over B, G a topological group, there is an action of G on E defined by right translation on each fiber. Now given F a space on which G acts on the left, we can define in a natural way a fiber bundle over B with fiber F. Indeed the bundle will be $(E \times F)/\sim$ where the equivalence relation $\sim$ is defined by identifying elements in $E \times F$ of the form $eg \times f$ and $e \times gf$ for $e \in E$, $f \in F$, $g \in G$. In the case of the Grassmannian above, we were taking $E = M^{reg}_{r,s}$, $B = Gr(r,s)$, $F = \mathbb{C}^r$, and $G = GL(r,\mathbb{C})$. For details about this construction see Steenrod [143].)

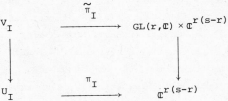

(iii) There is a simple geometric description of U_r and U_r^*. Indeed one defines a bundle (many times called the <u>universal subbundle</u>) $V \to Gr(r,s)$ to be the subbundle of $\mathbb{C}^s \times Gr(r,s)$ whose fiber at each point $\Lambda \in Gr(r,s)$ is the subspace $\Lambda \subseteq \mathbb{C}^s$. It is a simple exercise to show that V is in point of fact U_r^*. Now the quotient bundle $Q = (\mathbb{C}^s \times Gr(r,s))/V$ is called the <u>universal quotient bundle.</u> Via duality there is a natural identification $*: Gr(r,s) \to Gr(s-r,s)$ and the universal subbundle on $Gr(s-r,s)$ corresponds to the dual of the universal quotient bundle on $Gr(r,s)$ and similarly the universal quotient bundle on $Gr(s-r,s)$ corresponds to the dual of the universal subbundle in $Gr(r,s)$. In particular U_r is the pull-back under $*$ of the universal quotient bundle on $Gr(s-r,s)$.

(iv) We finally remark that the terminology "universal bundle" comes from the fact (for a proof see [44], page 104) that given any $\mathbb{C}^r$-vector bundle V over a variety S which is generated by global sections $\lambda_1,\dots,\lambda_s \in \Gamma(S,V)$ i.e. the fiber V_x is generated by $\lambda_1(x),\dots,\lambda_s(x)$ for each $x \in S$, there exists a morphism $\lambda: S \to Gr(r,s)$ such that $V \cong \lambda^*U_r$ and $\lambda_i = \lambda^*(u_i)$ where $u_1,\dots,u_s$ is a natural basis for $\Gamma(Gr(r,s),U_r)$ defined in terms of the transition data φ_I given above. (For the precise definitions see [44], page 101.) Conversely any morphism $\lambda: S \to Gr(r,s)$ induces a $\mathbb{C}^r$-vector bundle λ^*U_r generated by the global sections λ^*u_i. Note that in case $Gr(1,s) = \mathbb{P}^{s-1}$, we have $U_1 = \mathcal{O}_{\mathbb{P}^{s-1}}(1)$ (notation as in Part I (5.9)).

Before coming to the main result of this section we set some notation:

<u>Notation (5.2).</u>

(i) $V_{n,m,p} := \{(F,G,H) \in M_{n,n}(\mathbb{C}) \times M_{n,m}(\mathbb{C}) \times M_{p,n}(\mathbb{C}) \mid (F,G)$ is completely reachable$\}$.

(ii) $\mathcal{M}_{n,m,p} :=$ space of orbits defined by the action of $GL(n,\mathbb{C})$ on $V_{n,m,p}$ by change of basis in the state space $\mathbb{C}^n$ i.e. by $(F,G,H) \longmapsto (gFg^{-1}, gG, Hg^{-1})$ for $g \in GL(n,\mathbb{C})$.

We can thus state the following theorem [21], [57]:

<u>Theorem (5.3).</u>

(i) <u>The natural projection</u> $\tilde{u}_p: V_{n,m,p} \to V_{n,m}$ <u>descends to define</u> $u_p: \mathcal{M}_{n,m,p} \to \mathcal{M}_{n,m}$ <u>as a</u> $\mathbb{C}^{np}$<u>-vector bundle over</u> $\mathcal{M}_{n,m}$. <u>In particular</u> $\mathcal{M}_{n,m,p}$ <u>is a smooth algebraic variety.</u>

(ii) <u>For</u> $m > 1$, $p > 0$ <u>the bundles</u> $u_p: \mathcal{M}_{n,m,p} \to \mathcal{M}_{n,m}$ <u>are non-trivial.</u>

(iii) <u>The quotient morphism</u> $\pi_p: V_{n,m,p} \to \mathcal{M}_{n,m,p}$ <u>is a geometric quotient and defines</u> $V_{n,m,p}$ <u>as an algebraic principal</u> $GL(n,\mathbb{C})$<u>-bundle over</u> $\mathcal{M}_{n,m,p}$ <u>which is non-trivial for</u> $m > 1$. <u>In particular there exist no global algebraic canonical forms for</u> $m > 1$.

<u>Proof.</u> We prove the statements (i), (ii) and (iii) together. First to show that $u_p : \mathcal{M}_{n,m,p} \to \mathcal{M}_{n,m}$ is a $\mathbb{C}^{np}$-bundle is an easy exercise from our above constructions. Thus we have that $\mathcal{M}_{n,m,p}$ is a smooth algebraic variety fitting into a commutative diagram

$$\begin{array}{ccc} V_{n,m,p} & \xrightarrow{\tilde{u}_p} & V_{n,m} \\ \downarrow{\scriptstyle \pi_p} & & \downarrow{\scriptstyle \pi} \\ \mathcal{M}_{n,m,p} & \xrightarrow[u_p]{} & \mathcal{M}_{n,m} \end{array} .$$

Next in point of fact, the pre-stability of $V_{n,m}$ means that $V_{n,m} \to \mathcal{M}_{n,m}$ is a <u>universal</u> geometric quotient i.e. pull-backs (via the fiber product) are also geometric quotients. (See Mumford [104], pages 4 and 37.) Thus from the preceding diagram we conclude that $(\mathcal{M}_{n,m,p}, \pi_p)$ is a geometric quotient. Again by Mumford [104], page 16, this implies that $\pi_p : V_{n,m,p} \to \mathcal{M}_{n,m,p}$ is an algebraic principal $GL(n,\mathbb{C})$-bundle.

Now in order to show that such a principal bundle is non-trivial, it is equivalent to show that it has no global sections (Steenrod [143]). But taking the morphism φ as in Section 4, we get a morphism φ_p fitting into a commutative diagram

$$\begin{array}{ccc} \mathcal{M}_{n,m,p} & & \\ \downarrow{\scriptstyle u_p} & \searrow{\scriptstyle \varphi_p} & \\ \mathcal{M}_{n,m} & \xrightarrow[\varphi]{} & \mathbb{C}^n \end{array}$$

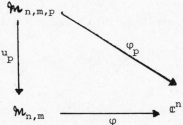

and from the projectivity of φ (4.4), we conclude that $\mathcal{M}_{n,m,p}$ is not quasi-affine when $m > 1$ and since $V_{n,m,p}$ is quasi-affine, there cannot exist any global section. In particular, for $m > 1$, there are no global algebraic canonical forms.

Finally by direct computation it is easy to verify that as vector bundles $\mathcal{M}_{n,m,p} \cong \mathcal{M}_{n,m,1} \oplus \cdots \oplus \mathcal{M}_{n,m,1}$ (the sum taken p times). Thus to prove (ii) we need only show that the $\mathbb{C}^n$-vector bundle $u_1 : \mathcal{M}_{n,m,1} \to \mathcal{M}_{n,m}$ is non-trivial. But again from the remarks of (5.1) (and using the same notation) it is immediate that $\bar{R}^* U_n$ is the $\mathbb{C}^n$-bundle associated to the principal $GL(n,\mathbb{C})$-bundle $V_{n,m} \to \mathcal{M}_{n,m}$.

Now from the definitions and constructions above it is easy to see that in point of fact $\mathcal{M}_{n,m,1}$ is also isomorphic to $\bar{R}^* U_n$, i.e. the $\mathcal{M}_{n,m,1}$ is the

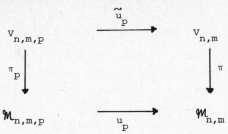

$\mathbb{C}^n$-bundle associated to the principal $GL(n,\mathbb{C})$-bundle $V_{n,m} \to \mathfrak{M}_{n,m}$. Thus the non-triviality of $V_{n,m} \to \mathfrak{M}_{n,m}$ for $m > 1$, implies (ii).

Q.E.D.

Remark (5.4). Mimicking the discussion in Section 3 (and actually Mumford's results in [108]), one can define a contravariant functor $\tilde{\mathfrak{F}}_p$: Var(k) $\to$ Sets by $\tilde{\mathfrak{F}}_p(S)$: = {isomorphism classes of families of completely reachable triples with m inputs, p outputs, and state space dimension n over S, i.e. isomorphism classes of $(m+p+2)$-tuples $(V, \tilde{F}, g_1,\ldots,g_m; h_1,\ldots,h_p)$ where $V, \tilde{F}, g_1,\ldots,g_m$ are as in (3.1) and $h_1,\ldots,h_p$ are global sections of the dual vector bundle $V^* \to S$} .

We leave it as an exercise to prove that $\mathfrak{M}_{n,m,p}$ represents $\tilde{\mathfrak{F}}_p$ or see e.g. Hazewinkel [54]. Also see our discussion in Part VI, Section 5.

§6. Some Open Problems

We would like to close this part of the notes with a discussion of some problems connected to the structure of $\mathfrak{M}_{n,m}$. The facts about the generic fibers of the morphism φ which we will discuss below were first told to the author by H. Kraft. The first problem is this:

Problem 1. Describe explicitly the fibers of the morphism $\varphi: \mathfrak{M}_{n,m} \to k^n$ (notation as in Section 4). Recall that we have a commutative diagram

$$\begin{array}{ccccc} V_{n,m} & \hookrightarrow & k^{n^2+nm} & \xrightarrow{\text{projection}} & k^{n^2} \\ \downarrow \pi & & & & \downarrow \alpha \\ \mathfrak{M}_{n,m} & & \xrightarrow{\quad\quad\varphi\quad\quad} & & k^n \end{array}$$

where α is the quotient morphism sending a matrix to its characteristic values described in Part III. Clearly $\pi^{-1} \cdot \varphi^{-1}(c_1,\ldots,c_n) = \{(F,G) \mid (F,G) \in V_{n,m}$ and the characteristic coefficients of the characteristic polynomial of F are $c_1,\ldots,c_n\}$. Moreover note that since k is algebraically closed the orbit of $O((F,G))$ always contains pairs $(\tilde{F},\tilde{G})$ with $\tilde{F}$ upper triangular.

Let $U \subset k^n$ be the dense subset consisting of images of matrices with distinct non-zero eigenvalues. Note that given a diagonal matrix F with distinct non-zero eigenvalues, that the stabilizer subgroup of F under conjugation consists of diagonal matrices with all diagonal elements non-zero. Since in $\varphi^{-1}(c)$, $c \in U$ we can always take representatives (F,G) with F diagonal, we may clearly identify $\varphi^{-1}(c)$ with $\{\text{stab}(F)G\}$ and the fact that G is part of a completely reachable pair, implies that we have that $\varphi^{-1}(c) \cong (\mathbb{P}^{m-1})^n$.

Now if we look at $\alpha^{-1}(c)$, $c \in k^n$, $\alpha^{-1}(c)$ will contain closures of orbits of $n \times n$ matrices under conjugation. (Recall from Part III, that unless the matrices are semisimple, these orbits are not in general closed.) In their beautiful paper [90], H. Kraft and C. Procesi study the closure of orbits of matrices under conjugation and conclude that the closures are normal, Cohen-Macaulay varieties with rational singularities. Looking at the preceding commutative diagram we see therefore that in $u: \mathcal{M}_{n,m} \to k^n$ we have a case in which products of projective spaces are degenerating to varieties with rational singularities. These degenerations should be very interesting from both a mathematical and system theoretic point of view.

Of course if one can solve Problem 1, we still have

Problem 2. Describe $\mathcal{M}_{n,m}$ as explicitly as possible. Closely related to the work of Kraft-Procesi [90] but apparently harder:

Problem 3. Describe the closure $\overline{O((F,G))}$. By the way, an easy exercise given our discussion in Part III, shows that the only closed orbits of pairs of matrices in k^{n^2+nm} consist of pairs of the form $(F_{ss},0)$ where F_{ss} is semisimple.

We also pose:

Problem 4. Find an explicit formula for $\dim \operatorname{stab}((F,G))$.

In part V, Section 3, we give a formula when $m = 1$.

Finally in the quotient space $\mathcal{M}_{n,m,p}$ of completely reachable triples, one can consider the Zariski open subset of canonical (i.e. completely reachable, completely observable) systems. Over the real numbers, when $m = p = 1$, this space (actually the space of proper transfer functions of MacMillan degree n) is disconnected and it is of great interest to study the topological properties of the connected components. We will return to this problem in Part VII of these notes.

PART V. LOCAL MODULI OF LINEAR TIME-INVARIANT DYNAMICAL SYSTEMS

In Part IV we studied the global moduli of finite dimensional constant linear dynamical systems using the techniques of geometric invariant theory. In particular (using the same notation as in Part IV) we have seen that for $m > 1$, the $GL(n,k)$ principal bundle $V_{n,m} \xrightarrow{\pi} \mathcal{M}_{n,m}$ admits no global sections and hence there cannot exist any global algebraic canonical forms. If $k = \mathbb{C}$, it is immediate that $V_{n,m} \xrightarrow{\pi} \mathcal{M}_{n,m}$ is a holomorphic non-trivial $GL(n,\mathbb{C})$ principal bundle for $m > 1$, and so there do not exist any global holomorphic canonical forms.

Now locally (in point of fact over the V_I, for I nice) one does have algebraic (and therefore holomorphic) sections i.e. there do exist algebraic local canonical forms. The problem we shall consider in this part of the lectures is to find a systematic method for constructing all such canonical forms even in the non-completely reachable case. This then will give us the local moduli of linear dynamical systems.

Just as the dominant theme of Part IV was the geometric invariant theory of Mumford here the essential techniques are due to Arnold. Arnold in [5] considers finding a holomorphic canonical form of $n \times n$ complex matrices under conjugation (for the exact definition see Section 1 below). We have already seen in Part III (2.5) that the Jordan canonical form isn't even continuous. We will see that Arnold's technique of constructing a holomorphic canonical form for matrices under conjugation immediately generalizes to pairs (F,G) and triples (F,G,H) of complex matrices of appropriate sizes corresponding to dynamical systems, acted on by the general linear group via change of basis in the state space (see Part IV). In point of fact we have already met such a holomorphic canonical form: The control canonical form of Part II, Section 4 (see Section 2 below). We shall actually concentrate on pairs of matrices and make some comments on triples in Section 4 (the techniques are identical).

Finally we remark that such holomorphic canonical forms are very important in system problems where there is parameter uncertainty and thus where reduction to a discontinuous canonical form can lead to a loss of information. We will discuss this more in Part VIII of the lectures. In this part we will use some standard terminology for the theory of complex manifolds and complex Lie groups. Good references for this are Morrow-Kodaira [103], and Wells [151]. The reader who is only interested in the algebraic case can consider all our complex manifolds to be smooth complex algebraic varieties, the holomorphic mappings to be morphisms, the complex Lie groups to be complex algebraic groups, and the term "holomorphic canonical form" can be replaced by "algebraic canonical form".

§1. Versal Deformations of Matrix Pairs

In this section we will use the idea of versality from deformation theory, to give precise meaning to the notion of "local holomorphic canonical form". We begin with the following general definition:

Definition (1.1). Let M be a complex manifold and let G be a complex Lie group acting on M. Let N be any complex manifold and $\varphi: N \to M$ a holomorphic mapping. Then we say that $\varphi: N \to M$ is versal at $n_o \in N$ if for every $\varphi': N' \to M$, N' a complex manifold, and φ' a holomorphic mapping such that for some $n_0' \in N'$, $\varphi'(n_0') = \varphi(n_o)$, there exists an open neighborhood U of n_o' in N', a holomorphic mapping $h: U \to N$ with $h(n_o') = n_o$, and a holomorphic mapping $\psi: U \to G$ with $\psi(n_o') = e$ (e is the identity element of G), such that $\varphi'|U = \psi\cdot(\varphi\circ h)$.

The condition of versality is very easy to check as the following theorem given by Arnold [5] for the case of $GL(n,\mathbb{C})$ acting on the space of $n \times n$ complex matrices $M_{n,n}(\mathbb{C})$ by similarity will reveal:

Theorem (1.2). Notation as in (1.1). Then $\varphi: N \to M$ is versal at $n_o \in N$ if and only if φ is transversal to the orbit of $\varphi(n_o)$ at n_o (that is, $T_{\varphi(n_o)}M = d\varphi T_{n_o}N + T_{\varphi(n_o)}(O(\varphi(n_o)))$, where $O(\varphi(n_o))$ = orbit of $\varphi(n_o)$ in M under the action of G, $d\varphi$ is the differential of φ, and if X is a complex manifold and $x \in X$, then T_xX denotes the tangent space of X at x).

Proof. The same method used by Arnold [5] (Lemma 2.3)) for the case of $GL(n,\mathbb{C})$ acting on $M_{n,n}(C)$ by similarity works in general. We will outline the argument in order to show it gives through under the hypotheses of (1.2).

Indeed if $\varphi: N \to M$ is versal at n_o, then for any $\varphi': N' \to M$ holomorphic with $\varphi'(n_o') = \varphi(n_o)$, there exists an open neighborhood U of n_o' in N', a holomorphic mapping $h: U \to N$, $h(n_o') = n_o$, and a holomorphic mapping $\psi: U \to G$ with $\psi(n_o') = e$ (the identity element of G) such that $\varphi'|U = \psi\cdot(\varphi\circ h)$. Let $\lambda: G \to M$ be defined by $\lambda(g) = g\cdot\varphi(n_o)$. Then we see easily that

$$d\varphi'_{n_o'} = d\lambda_e \circ d\psi_{n_o'} + d\varphi_{n_o} \circ dh_{n_o'} .$$

But this clearly implies φ must be transversal to the orbit of $\varphi(n_o)$ at n_o.

For the converse, let $\varphi: N \to M$ be transversal to the orbit of $\varphi(n_o)$ at n_o. Then replacing N by a submanifold if necessary, we may clearly assume that $\dim N = \dim M - \dim O(\varphi(n_o))$. Taking the differential of $\lambda: G \to M$ defined above gives a mapping $d\lambda_e: T_eG \to T_{\varphi(n_o)}M$ whose kernel is precisely the tangent space to $\mathrm{stab}(\varphi(n_o))$ (= stabilizer of $\varphi(n_o)$ in G). Let $S \subseteq G$ be a submanifold of complementary dimension to $\mathrm{stab}(\varphi(n_o))$ which is transversal to $\mathrm{stab}(\varphi(n_o))$ at the identity. Then if we define $u: S \times N \to M$ by $u(s,n) = s\cdot\varphi(n)$, it is easy to check $du_{(e,n_o)}$ is non-singular. (Note that $S \times N$ and M have the same dimension.) But by the inverse mapping theorem for holomorphic mappings (see [49], page 17), u is invertible on a sufficiently small neighborhood of (e,n_o) in $S \times N$. Next let $\varphi': N' \to M$, $\varphi'(n_o') = \varphi(n_o)$, be a holomorphic mapping. Then by the above for a sufficiently small neighborhood U of n_o', if $n' \in U$, we have $\varphi'(n') = u(s,n)$ for some $s \in S$, $n \in N$. Then if $p_1: S \times N \to S$ and $p_2: S \times N \to N$ are the projections, if we let $h: U \to N$ be defined by $h = p_2 \circ u^{-1} \circ \varphi'|U$, and $\psi: U \to G$ be defined by $\psi = p_1 \circ u^{-1} \circ \varphi'|U$, then $\varphi'|U = \psi\cdot(\varphi\circ h)$ showing φ is versal.

Q.E.D.

Definition (1.3). Let N be a complex manifold, $n_o \in N$ a base point, $\varphi: N \to M_{n,n}(\mathbb{C}) \times M_{n,m}(\mathbb{C})$ a holomorphic mapping. Set $(F(s),G(s)) := \varphi(s)$ for $s \in N$, and $(F,G) := (F(n_o),G(n_o))$. Then we call $\{(F(s),G(s))\}_{s\in N}$ a family of deformations of (F,G).

Remark (1.4). Now consider the action of $GL(n,\mathbb{C})$ on $M_{n,n}(C) \times M_{n,m}(\mathbb{C})$ given by $(F,G) \longmapsto (gFg^{-1},gG)$ for $(F,G) \in M_{n,n}(\mathbb{C}) \times M_{n,m}(\mathbb{C})$, $g \in GL(n,\mathbb{C})$. Let $\{(F(s),G(s))\}_{s\in N}$ be a family of deformations of (F,G) as in (1.3). Then from (1.1) we have a notion of versality for such families.

Suppose then $\{(F(s),G(s))\}_{s\in N}$ is a versal family of deformations of (F,G). This means that any family of pairs sufficiently close to (F,G) in C^{n^2+nm} can be reduced holomorphically to this versal family via the action of $GL(n,\mathbb{C})$.

Hence we make the following definition:

Definition (1.5). A local holomorphic canonical form of a given matrix pair $(F,G) \in M_{n,n}(\mathbb{C}) \times M_{n,m}(\mathbb{C})$ is a versal family of deformations of (F,G) over some base space N as in (1.4).

We conclude this section with some remarks which we will need in Sections 2 and 3:

Remarks (1.6).

(i) A versal family of deformations is, of course, not unique. However, from now on we make the convention that by "versal family" we mean "versal family of minimal dimension".

More precisely, if we have $\varphi: N \to M_{n,n}(\mathbb{C}) \times M_{n,m}(\mathbb{C})$ with $\varphi(n_o) = (F,G)$ a versal family of deformations of (F,G), then we require that

$$\dim N = n^2+nm-\dim O((F,G)) = nm+\dim \operatorname{stab}((F,G)) .$$

Note that in case (F,G) is completely reachable, $\dim \operatorname{stab}((F,G)) = 0$, and so $\dim N = nm$. This gives another proof that the moduli space of completely reachable pairs has dimension nm which we showed in Part IV (2.2). We consider N also to be a local moduli space for (F,G).

(ii) In the case of pairs of matrices, the condition of transversality of (1.2) is very easy to check. Indeed one need only note that the tangent vectors to the orbit of (F,G) in $M_{n,n}(\mathbb{C}) \times M_{n,m}(\mathbb{C}) \cong C^{n^2+nm}$, are those matrix pairs which can be represented in the form $([A,F],AG)$ for $A \in M_{n,n}(\mathbb{C})$ where $[A,F] = AF-FA$. See e.g. Humphreys [71].

(iii) Finally we remark that the condition of versality in (1.1) being local, means that it can be formulated in terms of germs of holomorphic mappings. Therefore, in particular, the local moduli space N may always be chosen to be some open neighborhood of the origin in some appropriate $\mathbb{C}^{\ell}$, and the base point $n_o \in N$ may be taken to be the origin.

§2. The Control Canonical Form

In this section we apply the ideas of Section 1 to the control canonical form which we will show is a local holomorphic canonical form for completely reachable matrix pairs, i.e. it defines a versal deformation. We rely on the treatments of Antoulas [3], Hazewinkel [55], Kalman [78], Popov [115]. We first recall some definitions from these references.

Definitions-Remarks (2.1).

(i) Let (F,G) be a completely reachable pair of matrices over an arbitrary field k with F $n\times n$, G $n\times m$. Denote the columns of G by $g_1,\dots,g_m$. We order the pairs of integers (i,j), $i = 0,\dots,n-1$, $j = 1,\dots,m$ lexicographically and denote this ordering by "$<$". Consider the following "Young's diagram":

	0	1	2		n-1
1	x			...	
2	x	x		...	
⋮	⋮	⋮	⋮		⋮
m	x	x	x		

where for each (i,j) we put in a cross if and only if the column vector $F^i g_j$ is linearly independent of the columns $F^{i'} g_{j'}$, for all $(i',j') < (i,j)$. Since the rank $(G\ FG \dots F^{n-1}G) = n$, we get in this way n crosses.

(ii) Let $\kappa_i :=$ number of crosses appearing in the i-th row of the diagram. Then $\kappa_1,\dots,\kappa_m$ are called the Kronecker indices associated to (F,G).

(iii) Let $\kappa =$ set of all (i,j) at which there appears a cross in the above Young's diagram. Then it is easy to see that if $(i,j) \in \kappa$, then $(i',j) \in \kappa$ for all $i' \leq i$ i.e. κ is a nice selection in the sense of Part IV (1.9). κ is called the Kronecker nice selection.

(iv) We let $I := \{j \mid 1 \leq j \leq m$ and such that a cross appears in the $(0,j)$ place of the Young's diagram associated to $(F,G)\}$. We will denote the elements of I by $j_1 \leq \dots \leq j_t$ where $t = \text{rank}\, G$. Notice that $\kappa_j \neq 0$ if and only if $j = j_r$ for some $1 \leq r \leq t$.

We can now state the following theorem which asserts the existence of the (flat) control canonical form:

Theorem (2.2). Notation as in (2.1). Then via change of basis in the state space, we may transform (F,G) to (F_*,G_*) of the following form:

$$F_* = ((F_*)_{ij}) \ , \quad i,j \in I : (F_*)_{ij} \in M_{\kappa_i \times \kappa_j}(k) \quad ;$$

$$(F_*)_{ii} = \begin{bmatrix} 0 & 1 & 0 & \cdots & 0 \\ 0 & 0 & 1 & \cdots & 0 \\ \vdots & & & & \\ -\alpha_{ii1} & \cdots\cdots & & & -\alpha_{ii\kappa_i} \end{bmatrix} , \quad (F_*)_{ij} = \begin{bmatrix} & 0 & \\ -\alpha_{ij1} & \cdots & -\alpha_{ij\kappa_j} \end{bmatrix}$$

for $i \neq j$, $\kappa_i > \kappa_j$;

$$(F_*)_{ij} = \begin{bmatrix} 0 & & & & & & \\ -\alpha_{ij1} & \cdots & -\alpha_{ij\kappa_i} & 0 & \cdots & 0 \end{bmatrix}$$

for $i \neq j$, $\kappa_i \leq \kappa_j$.

$$G = ((G_*)_{ij}) \ , \quad i \in I \ , \quad 1 \leq j \leq m \ , \quad (G_*)_{ij} \in M_{\kappa_i \times 1}(k) \quad ;$$

$$(G_*)_{ii} = \begin{pmatrix} 0 \\ \vdots \\ 0 \\ 1 \end{pmatrix} \quad i \in I \ ; \quad (G_*)_{ij} = \begin{pmatrix} 0 \\ \vdots \\ -\alpha_{ij(\kappa_j+1)} \end{pmatrix}$$

for $i \neq j \in I$, $\kappa_i > \kappa_j$;

$$(G_*)_{ij} = \begin{pmatrix} 0 \\ \vdots \\ 0 \end{pmatrix} \quad \text{for } i \neq j \in I \ ; \ \kappa_i \leq \kappa_j \quad ;$$

$$(G_*)_{ij} = \begin{pmatrix} 0 \\ \vdots \\ -\alpha_{ij1} \end{pmatrix} \quad j \notin I \ .$$

Proof. For the definition of the α_{ijk} and the proof of the theorem see either Antoulas [3], Hazewinkel [55], or Popov [115].

Q.E.D.

We can now prove the main theorem of this section. Note we once again work over the complex numbers $\mathbb{C}$ in what follows below.

Theorem (2.3). The control canonical form of (2.2) is a local holomorphic canonical form.

Proof. We will prove this theorem for the case of a completely reachable (F,G)

with $n = 3$, $m = 2$, $\kappa_1 = 2$, $\kappa_2 = 1$. The proof of the general case is the same but we want to keep the notation as simple as possible to illustrate the ideas of Section 1.

With these assumptions via the action of $GL(n,\mathbb{C})$ we can reduce (F,G) to matrices of the form

$$F_* = \begin{bmatrix} 0 & 1 & 0 \\ a_1 & a_2 & a_3 \\ a_4 & 0 & a_5 \end{bmatrix}, \quad G_* = \begin{bmatrix} 0 & 0 \\ 1 & a_6 \\ 0 & 1 \end{bmatrix}$$

by Theorem (2.2).

Now we define a holomorphic mapping from $\mathbb{C}^6 \xrightarrow{\varphi} M_{3,3}(\mathbb{C}) \times M_{3,2}(\mathbb{C})$ by

$$(s_1,s_2,s_3,s_4,s_5,s_6) \longmapsto \left(\begin{bmatrix} 0 & 1 & 0 \\ a_1+s_1 & a_2+s_2 & a_3+s_3 \\ a_4+s_4 & 0 & a_5+s_5 \end{bmatrix}, \begin{bmatrix} 0 & 0 \\ 1 & a_6+s_6 \\ 0 & 1 \end{bmatrix} \right).$$

From Section 1 to show that the control canonical form is holomorphic i.e. that pairs of matrices sufficiently close to (F,G) can be brought to control canonical form without losing their holomorphic dependence on the parameters, we must show that the family $(F_*(s),G_*(s)) := \varphi(s)$ is a versal family of deformations of $(F,G) = \varphi(0)$ and thus by Theorem (1.2) we must show that φ is transversal to the orbit of (F,G) at 0.

To carry out this computation we first note that without loss of generality we may assume that $a_1 = a_2 = \ldots = a_6 = 0$. Next we identify $M_{3,3}(\mathbb{C}) \times M_{3,2}(\mathbb{C})$ with $\mathbb{C}^{15}$ in the obvious way, and then we may identify the tangent space to (F,G) in $\mathbb{C}^{15}$ with $\mathbb{C}^{15}$ itself. Moreover we can identify the tangent space of $\mathbb{C}^6$ at the origin with $\mathbb{C}^6$ itself.

With these assumptions and identifications, computing the differential of φ at the origin and using (1.6) (ii), it is easy to see that we must show that

$$(*) \qquad ([A,F_*(0)],AG_*(0)) + \left(\begin{bmatrix} 0 & 0 & 0 \\ s_1 & s_2 & s_3 \\ s_4 & 0 & s_5 \end{bmatrix}, \begin{bmatrix} 0 & 0 \\ 0 & s_6 \\ 0 & 0 \end{bmatrix} \right)$$

generates $\mathbb{C}^{15}$ as A varies in $M_{3,3}(\mathbb{C})$ and $(s_1,\ldots,s_6)$ varies in $\mathbb{C}^6$. Let

$$A = \begin{bmatrix} b_1 & b_2 & b_3 \\ b_4 & b_5 & b_6 \\ b_7 & b_8 & b_9 \end{bmatrix}.$$

Then

$$[A,F_*(0)] + \begin{bmatrix} 0 & 0 & 0 \\ s_1 & s_2 & s_3 \\ s_4 & 0 & s_5 \end{bmatrix} = \begin{bmatrix} -b_4 & b_1-b_5 & -b_6 \\ s_1 & s_2+b_4 & s_3 \\ s_4 & b_7 & s_5 \end{bmatrix}$$

and

$$AG_*(0) + \begin{bmatrix} 0 & 0 \\ 0 & s_6 \\ 0 & 0 \end{bmatrix} = \begin{bmatrix} b_2 & b_3 \\ b_5 & b_6+s_6 \\ b_8 & b_9 \end{bmatrix} .$$

Hence it is clear (*) generates $\mathbb{C}^{15}$ as A varies in $M_{3,3}(\mathbb{C})$ and $(s_1,\ldots,s_6)$ varies in $\mathbb{C}^6$.

Q.E.D.

Remark (2.4). In Casti [22], pages 104-105, a representation of the control canonical form is given which in the case $n = 3$, $m = 2$, $\kappa_1 = 2$, $\kappa_2 = 1$ would give that

$$F_* = \begin{bmatrix} 0 & 1 & 0 \\ a_1 & a_2 & a_3 \\ a_4 & a_5 & a_6 \end{bmatrix} , \quad G_* = \begin{bmatrix} 0 & 0 \\ 1 & 0 \\ 0 & 1 \end{bmatrix}$$

that is all the parameters are contained in F_*.

We have seen from (2.2) this form is incorrect, but it is an interesting exercise to detect the mistake using the deformation theory from Section 1. In point of fact, a computation analogous to that carried out in (2.3) shows that the corresponding family associated to (F_*,G_*) is not versal and hence this erroneous canonical form is not a local holomorphic canonical form.

§3. On the Construction of Holomorphic Canonical Forms

In Section 2 we have applied the theory of Section 1 to the completely reachable case. However in point of fact the ideas of versality are independent of the property of complete reachability and may be used to derive holomorphic canonical forms in the general case. This will be our project here. We first start with an illustrative example, generalizing an example of Arnold's [5] (Corollary (3.4)).

Example (3.1). Let $F = J(\lambda;n)$ be an $n \times n$ Jordan block with eigenvalue λ and let $G = e_k^t$ be the transpose of the k-th unit basis vector in $\mathbb{C}^n$. Then using (1.2) and (1.5) (ii) it is easy to check that the family

$$\lambda I + \begin{bmatrix} 0 & 1 & 0 & \ldots & 0 \\ 0 & 0 & 1 & \ldots & 0 \\ . & . & . & \ldots & . \\ \alpha_1 & . & . & \ldots & \alpha_n \end{bmatrix} , \quad G + \begin{bmatrix} 0 \\ \vdots \\ 0 \\ \beta_1 \\ \vdots \\ \beta_{n-k} \end{bmatrix}$$

where I is the $n \times n$ identity matrix and the α_i and β_j are independent parameters, determines a (2n-k)-dimensional versal deformation of (F,G).

We can get a uniform procedure for determining versal deformations and thereby defining holomorphic canonical forms via the following theorem:

Theorem (3.2). Let $(F,G) \in M_{n,n}(\mathbb{C}) \times M_{n,m}(\mathbb{C})$. Let $S: = \{(B,C) \in M_{n,n}(\mathbb{C}) \times M_{n,m}(\mathbb{C}) \mid [F,\bar{B}^t] + G\bar{C}^t = 0\}$. Then the family $\{(F+B,\ G+C)\}_{(B,C) \in S}$ gives a versal family of deformations of (F,G) . In particular

$$\dim S = nm + \dim \operatorname{stab}((F,G)) .$$

Proof. We use a similar method as that of Arnold [5] ((4.1) and (4.2)). Accordingly we define an Hermitian scalar product on $M_{n,n}(\mathbb{C}) \times M_{n,m}(\mathbb{C})$ by

$$(F_1,G_1)\cdot(F_2,G_2): = \operatorname{Tr}(F_1\bar{F}_2^t) + \operatorname{Tr}(G_1\bar{G}_2^t)$$

where "Tr" stands for "trace". Now the tangent vectors to the orbit of (F,G) are the pairs of matrices that can be represented in the form $([A,F],AG)$ where $A \in M_{n,n}(\mathbb{C})$. Then relative to the above Hermitian product, (B,C) is perpendicular to the orbit of (F,G) if and only if for every $A \in M_{n,n}(\mathbb{C})$ we have

$$(*) \qquad \operatorname{Tr}([A,F]\bar{B}^t) + \operatorname{Tr}(AG\bar{C}^t) = 0 .$$

But

$$\operatorname{Tr}([A,F]\bar{B}^t) + \operatorname{Tr}(AG\bar{C}^t) = \operatorname{Tr}(A([F,\bar{B}^t] + G\bar{C}^t))$$

and this implies that (*) holds for all $A \in M_{n,n}(\mathbb{C})$ if and only if $[F,\bar{B}^t] + G\bar{C}^t = 0$. Then by (1.2) we see that $\{(F+B,G+C)\}_{(B,C) \in S}$ is a versal family of deformations of (F,G) .

Finally the fact that $\dim S = nm + \dim \operatorname{stab}((F,G))$ is now an immediate corollary of (1.6) (i).

Q.E.D.

Remarks (3.3).

(i) From (3.2) we can now compute a versal deformation of any given pair of matrices (F,G) and hence determine a holomorphic canonical form. In point of fact Theorem (3.2) gives us a set of linear equations which determine the versal deformation and hence give equations for the local moduli space of matrix pairs. In case $m = 1$, one can start from a discontinuous canonical form due to Byrnes-Gauger [20] which generalizes the Lure-Lefschetz-Letov form [22], in order to write down simple equations for the local moduli space. This seems of interest and so we will specifically carry out this procedure now.

Explicitly let $(F,G) \in M_{n,n}(\mathbb{C}) \times M_{n,1}(\mathbb{C})$, and suppose that F has eigenvalues $\lambda_1,\ldots,\lambda_r$ with the λ_i-eigenspace being of dimension n_i , $1 \leq i \leq r$. Then via change of basis in the state space from [20], page 86, we see that (F,G) is equivalent to $(F_J,\tilde{G})$ where F_J is the Jordan canonical form of F and

$$\tilde{G} = \begin{bmatrix} g_1 \\ g_2 \\ \vdots \\ g_r \end{bmatrix}$$

where g_i is $n_i \times 1$ $(1 \leq i \leq r)$ and has at most one non-zero entry which is a 1.

Therefore to compute a versal deformation space of $(F,G) \in M_{n,n}(\mathbb{C}) \times M_{n,1}(\mathbb{C})$ we can always assume that it has the canonical form described above. Moreover to write down equations for such a versal space using (3.2) we can easily reduce ourselves to the case in which F has one eigenvalue which we may take to be zero. Then if $F = J(0;n_r) \oplus \dots \oplus J(0;n_1)$, $\sum_{i=1}^{r} n_i = n$, $r \geq 2$, $n_r \geq \dots \geq n_1$, and $G = e_k^t$ where e_k is the standard unit basis vector in $\mathbb{C}^n$, for the solution of $[F,A] + GT = 0$ (see (3.2)) where $A = (a_{ij})$ is $n \times n$ and $T = (t_1,\dots,t_n)$, we have the following (not necessarily independent) system of equations.

(1) $a_{k,j-1} - a_{k+1,j} + t_j = 0$ for

$k \neq n_r, n_r+n_{r-1},\dots,n_r+n_{r-1}+\dots+n_1 = n$

$j \neq 1, n_r+1, n_r+n_{r-1}+1,\dots,n_r+n_{r-1}+\dots+n_2+1$.

(2) $a_{k+1,\ell} = t_\ell$ for $\ell = 1, n_r+1,\dots,n_r+n_{r-1}+\dots+n_2+1$
$k \neq n_r, n_r+n_{r-1},\dots,n$.

(3) $a_{i,j-1} = a_{i+1,j}$ for $i \neq k, n_r, n_r+n_{r-1},\dots,n$
$j \neq 1, n_r+1,\dots,n_r+n_{r-1}+\dots+n_2+1$.

(4) $a_{i+1,\ell} = 0$ for $\ell = 1, n_r+1,\dots,n_r+n_{r-1}+\dots+n_2+1$
$i \neq k, n_r, n_r+n_{r-1},\dots,n$.

(5) $a_{u,j-1} = 0$ for $k \neq n_r, n_r+n_{r-1},\dots,n$
$u = n_r, n_r+n_{r-1},\dots,n$
$j \neq 1, n_r+1,\dots,n_r+n_{r-1}+\dots+n_2+1$.

(6) $a_{k,j-1} = -t_j$ for $k = n_r, n_r+n_{r-1},\dots,n$
$j \neq 1, n_r+1,\dots,n_r+n_{r-1}+\dots+n_2+1$.

(ii) We also note that the canonical form of (i) gives a simple formula for $\dim \operatorname{stab}(F,G))$ in the scalar input case and hence a simple formula for the dimension of the local moduli space via (1.6) (i).

Indeed we assume that (F,G) has the above canonical form and again we are easily reduced to the case in which F has only one eigenvalue, say $F = J(\lambda;n_r) \oplus \dots \oplus J(\lambda;n_1)$ where $\sum_{i=1}^{r} n_i = n$, $n_r \geq \dots \geq n_1$. Let $q = \operatorname{rank}(G\ FG \dots F^{n-1}G)$ and suppose that if we identify $\mathbb{C}^n$ with $\mathbb{C}^{n_r} \oplus \dots \oplus \mathbb{C}^{n_1}$ so that we may write $G = G_r + \dots + G_1$ where $G_i \in C^{n_i}$, if $G \neq 0$, let j be such that G_j contains the only non-zero entry 1 of G. Then we claim that $\dim \operatorname{stab}((F,G)) = n_r + 3n_{r-1} + \dots + (2r-1)n_1 - qr +$ (number of i such that $i < j$ and $n_i < n_j$).

To see this, first note that $\dim \mathrm{stab}((F,G)) = \dim \mathrm{stab}(F) - \dim(\mathrm{stab}(F))G$. But from Gantmacher [39], Vol. I, page 222

$$\dim \mathrm{stab}(F) = n_r + 3n_{r-1} + \ldots + (2r-1)n_1 .$$

Thus we must calculate $\dim \mathrm{stab}(F)G$. But the group $\mathrm{stab}(F)$ is well-known (see Gantmacher [39], Vol.I, pages 220-221) and hence from the specific form of the elements of $\mathrm{stab}(F)$ it is easy to show $\dim \mathrm{stab}(F)G = q\cdot r$ - (number of i such that $i < j$ and $n_i < n_j$) giving us the formula.

We conclude with the following example exhibiting the techniques of (3.2) and (3.3):

Example (3.4). Let

$$F = \begin{bmatrix} 0 & 1 & 0 & 0 & 0 \\ 0 & 0 & 0 & 0 & 0 \\ 0 & 0 & 0 & 1 & 0 \\ 0 & 0 & 0 & 0 & 0 \\ 0 & 0 & 0 & 0 & 0 \end{bmatrix}, \quad G = \begin{bmatrix} 1 \\ 0 \\ 0 \\ 0 \\ 0 \end{bmatrix} .$$

Then using the equations of (3.3) (i) or by direct computation the family

$$F(t) = \begin{bmatrix} t_1 & 1 & t_2 & 0 & 0 \\ t_3 & t_4 & t_5 & t_2 & t_6 \\ t_7 & 0 & t_8 & 1 & 0 \\ t_9 & t_{10} & t_{11} & t_8 & t_{12} \\ t_{13} & t_{14} & t_{15} & 0 & t_{16} \end{bmatrix}, \quad G(t) = \begin{bmatrix} 1 \\ t_4 - t_1 \\ 0 \\ t_{10} - t_7 \\ t_{14} \end{bmatrix}$$

defines a versal 16 parameter deformation of (F,G). Hence again we have a local holomorphic canonical form in the sense that any family of pairs sufficiently close to (F,G) can be reduced holomorphically to $(F(t),G(t))$.

§4. Versal Deformations of Triples of Matrices

We would like to conclude Part V with some brief comments about triples of matrices. We let $GL(n,\mathbb{C})$ act on $M_{n,n}(\mathbb{C}) \times M_{n,m}(\mathbb{C}) \times M_{p,n}(\mathbb{C})$ in the standard way. Theorem (1.1) of course applies to this situation and we can write a formula for the dimension of a local moduli space N for a triple (F,G,H), namely

$$\dim N = nm + np + \dim \mathrm{stab}((F,G,H)) .$$

As for writing down local holomorphic canonical forms explicitly we have the following result trivially generalizing (3.2) :

Theorem (4.1). *Let* $(F,G,H) \in M_{n,n}(\mathbb{C}) \times M_{n,m}(\mathbb{C}) \times M_{p,n}(\mathbb{C})$. *Let* $T: = \{(B,C,D) \in M_{n,n}(\mathbb{C}) \times M_{n,m}(\mathbb{C}) \times M_{p.n}(\mathbb{C}) \mid [F,\bar{B}^t] + G\bar{C}^t - H\bar{D}^t = 0\}$. *Then the family* $\{(F+B,G+C,H+D)\}_{(B,C,D)\in T}$ *gives a versal family of deformations of* (F,G,H) *and in particular*

$$\dim T = nm + np + \dim \mathrm{stab}((F,G,H)) .$$

<u>Proof.</u> Just define an Hermitian inner product on $M_{n,n}(\mathbb{C}) \times M_{n,m}(\mathbb{C}) \times M_{p,n}(\mathbb{C})$ by $(F_1,G_1,H_1)\cdot(F_2,G_2,H_2) := \mathrm{Tr}(F_1\bar{F}_2^t) + \mathrm{Tr}(G_1\bar{G}_2^t) + \mathrm{Tr}(H_1\bar{H}_2^t)$. The tangent vectors to the orbit of (F,G,H) are triples of the form $([A,F],AG,-HA)$ where $A \in M_{n,n}(\mathbb{C})$. The proof then proceeds as in (3.2).

Q.E.D.

We leave it as an exercise for the interested reader to write down some examples of versal deformations of triples of matrices.

PART VI. ALGEBRAIC REALIZATION THEORY

We recall briefly and informally the realization problem discussed in Part II, Section 1. Accordingly we are given a "black box" Σ having m input and p output terminals. We imagine that at time t the black box acts on a given input signal and one observes a response at the output terminals. The realization problem is then to model Σ by a dynamical system in the sense of Part II (1.1).

In case we wish to model Σ by a finite dimensional linear dynamical system, a complete solution has been given by Kalman [82], [75]. The case of realizations of linear systems over rings has been studied by Byrnes [17], Eilenberg [28], Rouchaleau-Wyman [121], Sontag [137], and many others. A study of the realization of polynomial systems was carried out in the monograph of Sontag [139]. Finally realizations of transfer functions using polynomial methods has been carried out by Fuhrmann [35]. We shall discuss some of the work of these mathematicians below.

We might add that one of the fundamental motivations for the study of the geometry of rational transfer functions which will be the topic of Part VII of these lectures, comes from the so-called partial realization problem [81]. This too will be touched upon in Section 4 below.

§1. Input/Output Maps and Abstract Realization Theory

This section contains the fundamental ideas upon which modern algebraic realization theory is based. The point of view taken here is due to Kalman [82] (Chapter 10) and [75].

Throughout this section Σ will denote a linear time-invariant finite dimensional discrete dynamical system defined over an arbitrary field k. We adopt the notation from Part II, Section 1. In particular we let $U \cong k^m$ be the set of input values, $Y \cong k^p$ the set of output values, $X \cong k^n$ the state space, Ω = input space, Γ = output space, φ = the state transition map, η = the readout map.

Now recall from Part II, the idea of realization theory is to go from an external description of a system given by an input/output map to an internal state space description. We will therefore need a definition of "input/output map" which corresponds to the kinds of systems we have in mind. As noticed by Kalman [82] (Chapter 10), the input/output maps we need should formalize the notion of an experiment gotten by applying an input sequence of finite duration and then observing the output after the input is stopped. Since we are only interested in constant systems we can always assume that the input sequences are terminated at time $t = 0$, at which time we begin to observe the output sequences. The exact definitions are:

Definitions (1.1).

(i) We set

$\Omega := \{u: \mathbb{Z} \to U \mid u(t) = 0$ for every $t > 0$ and such that there exists $t_0 \leqslant 0$ (which may depend on u) such that $u(t) = 0$ for every $t < t_0\}$. The length of $u \in \Omega$, denoted $|u|$, is then $\max\{-t \in \mathbb{Z} \mid u(\tau) = 0$ for $\tau < t\}$.

(ii) $\Gamma := \{\gamma: \mathbb{Z} \to Y \mid \gamma(t) = 0$ for every $t \leqslant 0\}$.

(iii) We define the shift operator $\sigma_\Omega: \Omega \to \Omega$ by $\sigma_\Omega u(t) := u(t+1)$ and the shift operator $\sigma_\Gamma: \Gamma \to \Gamma$ by $\sigma_\Gamma \gamma(t) := \gamma(t+1)$ for $t > 0$, $\sigma_\Gamma \gamma(t) = 0$ for $t \leqslant 0$.

(iv) A discrete-time constant input/output map is a map $f: \Omega \to \Gamma$ such that the following diagram is commutative:

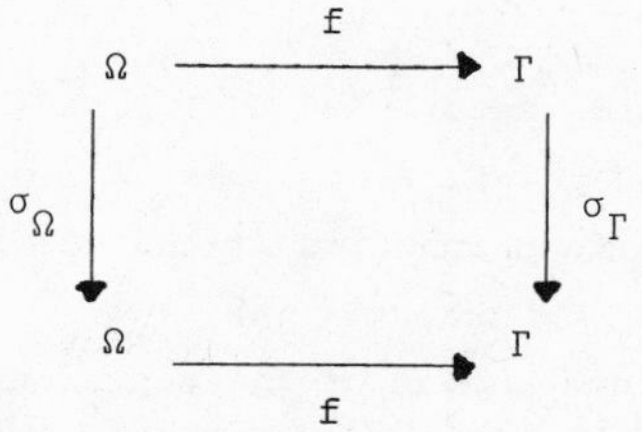

f is linear if it is linear with respect to the natural k-vector space structures on Ω and Γ. (These structures are the obvious ones.)

Definitions-Remarks (1.2).

(i) Given our discrete time system Σ above, if we take equilibrium state 0 (i.e. a state in which the system remains until the application of inputs u), then we get an input/output map $f_\Sigma: \Omega \to \Gamma$ defined by .

$$(f_\Sigma(u))(t) := \begin{cases} \eta(\varphi(0; -|\sigma_\Omega^{t-1} u|, 0, \sigma_\Omega^{t-1} u)) & \text{for } t > 0; \\ 0 & \text{otherwise} \end{cases}$$

where σ_Ω^{t-1} := composition of σ_Ω with itself taken $t-1$ times.

(ii) Σ is a realization of f a linear constant input/output map of $f = f_\Sigma$.

The key step in any such realization problem is to define the state space from a given f. For our set-up here this is actually quite trivial on the abstract set theoretical level and the main idea was already given by Nerode [111] (see also Part II (1.4)). This solution will be very important for us and so we would like to discuss it here. This leads us to the following:

Nerode Equivalence (1.3).

(i) We use the notation of (1.1). For $u \in \Omega$, let I_u := support of u := $\{t \in \mathbb{Z} \mid u(t) \neq 0\}$.

(ii) For $u, u' \in \Omega$ such that $I_u \cap I_{u'} = \emptyset$, we set

$$u+u' := \begin{cases} u & \text{on } I_u \\ u' & \text{on } I_{u'} \end{cases} .$$

(iii) We define the operation $\circ: \Omega \times \Omega \to \Omega$ of <u>concatenation</u> by $u \circ u' := \sigma_\Omega^{|u'|} u + u'$ where $\sigma_\Omega^{|u'|}$:= composition of σ_Ω with itself taken $|u'|$ times. With the operation of concatenation it is clear that Ω is a monoid with the neutral element $0: \mathbb{Z} \to U$ being defined by $0(t) \equiv 0$.

(iv) Let f be a discrete constant input/output map. Then we say that $u, u' \in \Omega$ are <u>Nerode equivalent with respect to</u> f , denoted by $u \underset{f}{\sim} u'$, if

$$f(u \circ u'') = f(u \circ u'') \quad \text{for all} \quad u'' \in \Omega .$$

We denote the equivalence class containing $u \in \Omega$ by $(u)_f$, and the set of all equivalence classes by X_f . We then have :

<u>Proposition (1.4). Let</u> f <u>be a linear constant discrete-time input/output map. Then there exists a realization</u> Σ_f <u>for</u> f .

<u>Proof.</u> We leave the complete details of the proof as an easy exercise or see [75], page 41. The main idea of course is to take X_f of (1.3) (iv) to be the state space of Σ_f . The state transition function is then defined by

$$\varphi(0; -|u'|, x, u') := (u \circ u')_f \quad \text{where} \quad x = (u)_f .$$

(It is immediate that this is independent of the representative u of x .) The readout map $\eta: X_f \to Y$ is given by $\eta(x) := f(u)(1)$ (where $x = (u)_f$). Again it is easy to show this is well-defined.

Q.E.D.

With this formalism out of the way we can now give:

<u>Kalman Realization (1.5).</u> Following our previous conventions we set $U := k^m$ and $Y := k^p$. f will denote a linear constant discrete-time input-output map throughout this discussion.

Motivated by the notion of Laplace transform (Part II (5.1); see also [82], page 245) given an alement $u \in \Omega$ ((1.1)), one can identify u with the formal <u>finite</u> sum $\sum_{t \in \mathbb{Z}} u(t) z^{-t}$ where z is some indeterminate. Thus with this identification, one has an isomorphism of k-vector spaces $\Omega \cong k^m[z]$:= the vector space of polynomials in z with coefficients in k^m .

Now $k^m[z]$ is a $k[z]$-module in the obvious way and thus via the preceding isomorphism, we can induce a $k[z]$-module structure on Ω . Note, and this is crucial, under this isomorphism the shift operator $\sigma_\Omega: \Omega \to \Omega$ becomes identified with multiplication by z in the $k[z]$-module $k^m[z]$.

As for Γ , given $\gamma \in \Gamma$, again following the above Laplace transform ideas, we identify γ with the (<u>not</u> necessarily finite) formal Laurent series $\sum_{t \in \mathbb{Z}} \gamma(t) z^{-t}$. Since by definition $\gamma(t) = 0$ for $t \leq 0$, we have that under this identification an isomorphism of k-vector spaces $\Gamma \cong z^{-1} k^p[[z^{-1}]]$ where $k^p[[z^{-1}]]$:= formal power series in z^{-1} with coefficients in k^p . With this identification one can induce a $k[z]$-module structure on Γ by setting for $p \in k[z]$, $p \cdot \gamma := p(\sigma_\Gamma)\gamma$ = the power series gotten by multiplying p and γ in the ordinary way and then truncating the non-negative terms from the resulting product. (Here $\sigma_\Gamma: \Gamma \to \Gamma$ is the shift operator.) From the commutative diagram in the definition of f (1.1)(iv) and from the fact σ_Ω can be identified with multiplication by z, relative to the above $k[z]$-module structures on Ω and Γ, $f: \Omega \to \Gamma$ becomes a $k[z]$-module homomorphism.

We are now ready to realize f in an interesting way, i.e. by a finite dimensional canonical linear system Σ_f. We will take the state space of our realization of f to be $\Omega/\ker f$. The beautiful fact is that the equivalence relation defined in this way (i.e. u is equivalent to u' just in case $f(u) = f(u')$) is precisely the Nerode equivalence of (1.3). Indeed we must show that $f(u \circ u'') = f(u' \circ u'')$ for all $u'' \in \Omega$ if and only if $f(u) = f(u')$. But $u \circ u'' = \sigma_\Omega^{|u''|} u + u'' = z^{|u''|} u + u''$ so that $f(z^{|u''|} u + u'') = f(z^{|u''|} u' + u'')$ for all u'' if and only if $f(u) = f(u')$ since f is a $k[z]$-module homomorphism. Hence the Kalman state space $\Omega/\ker f$ is essentially the Nerode state space X_f but now with the extra $k[z]$-module structure. Of course we will take Ω as our input space, Γ as our output space for the realization Σ. Let us set $X := \Omega/\ker f$, and we assume that X has dimension n.

Next recall from Part II that in this linear discrete constant finite dimensional case, the system Σ will be defined by a triple of matrices over k (F,G,H) where F is $n \times n$, G is $n \times m$, and H is $p \times n$. In these terms then the state transition map $\varphi: \Gamma \times \Gamma \times X \times \Omega \to X$ is given by $\varphi(t+1;t,x,u) = Fx(t) + Gu(t)$ and the readout map $\eta: X \to Y$ is given by $\eta(x) = Hx$. Thus we write our system in the form

$$x(t+1) = Fx(t) + Gu(t)$$

$$y(t) = Hx(t) .$$

(In the constant case the readout map is independent of t.)

We accordingly define now

$\underline{F}_f: X \to X$ by $x \longmapsto z \cdot x$,

$\underline{G}_f: \Omega \to X$ by $u \longmapsto u \bmod \ker f$, and

$\underline{H}_f: X \to \Gamma$ by $u \bmod \ker f \longmapsto f(u)$.

Since we have identified Ω with $k^m[z]$ we can restrict $\bar{G}_f$ to k^m and thus get a k-linear map $G: k^m \to X$. Moreover $\bar{H}_f: X \to \Gamma$ must be such that $(\bar{H}_f X)(t) = H_f z^{t-1} x$ for $t > 0$ for some k-linear $H_f: X \to k^p$. Then the <u>Kalman realization</u> of f is $\Sigma_f := (F_f, G_f, H_f)$.

<u>Theorem (1.6).</u> <u>Notation as in</u> (1.5). <u>The system</u> Σ_f <u>is a canonical realization of the input/output map</u> f.

<u>Proof.</u> We first show that Σ_f indeed realizes f. Now in general given a system $\Sigma = (F,G,H)$ it is immediate from the definitions (see (1.2)(i)) that the input/output map $f_\Sigma: \Omega \to \Gamma$ is defined by

$$(f_\Sigma(u))(\tau) := \begin{cases} \sum_{t\in\mathbb{Z}} H F^{-t+\tau-1} Gu(t) & \text{for } \tau > 0 ; \\ 0 & \text{for } \tau \leq 0 . \end{cases}$$

But then a simple computation shows that $f_{\Sigma_f} = \bar{H}_f \circ \bar{G}_f$. By construction however $f = \bar{H}_f \circ \bar{G}_f$ so that Σ_f realizes f .

Next using the facts that $F_f \colon X \to X$ corresponds to multiplication by z , that $\bar{H}_f$ and $\bar{G}_f$ are $k[z]$-module homomorphisms, and finally the results from Part II (2.3) and (3.4), it is clear that Σ_f is completely reachable if and only if $\bar{G}_f$ is surjective and Σ_f is completely observable if and only if $\bar{H}_f$ is injective. But this is trivial from the definitions of $\bar{G}_f$ and $\bar{H}_f$.

Q.E.D.

Remarks (1.7).

(i) The importance of the Kalman realization aside from explicitly using the natural $k[z]$-module structure on the state space to construct a realization, is that his realization is canonical. Indeed if the problem were only to realize f we could take the state space to be Ω , F_f to be σ_Ω , $\bar{G}_f$ the identity, and $\bar{H}_f$ to be f itself.

(ii) The discussion above shows that given constant linear input/output map $f\colon \Omega \to \Gamma$, that we may equivalently define a realization of f to be a factorization of f as

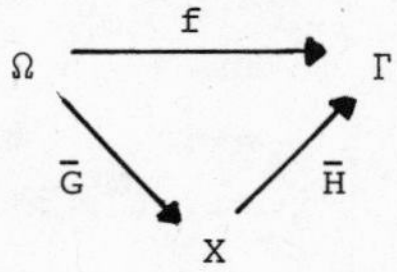

where X is a $k[z]$-module and $\bar{G}$ and $\bar{H}$ are $k[z]$-module homomorphisms (f is automatically a $k[z]$-homomorphism by definition). Then the realization will be completely reachable if and only if G is surjective, and completely observable if and only if $\bar{H}$ is injective.

With this description of a realization, by some abstract nonsense (see Kalman [82], pages 258-259) it is easy to show that any two canonical realizations of f are isomorphic.

(iii) Finally it is a simple exercise to show ([82] (6.10), page 259) that a canonical realization of f is minimal in the sense that the state space has minimal dimension among the dimensions of the state spaces of all other realizations. Conversely, a minimal realization is also canonical. We will see in Section 5 that the situation for systems over rings is much more complicated.

§2. Hankel Matrices

In this section we want to mention a few specific algorithms for constructing realizations as well as giving simple conditions when finite dimensional realizations exist. The key object will be the classical "Hankel matrix" (to be defined below). The Hankel matrix will also be important in deducing an interesting property concerning the moduli of linear dynamical systems. Throughout this section we let $f: \Omega \to \Gamma$ be a linear constant input/output map exactly as in Section 1. In particular we have the space of input values $U \cong k^m$, and output values $Y \cong k^p$.

Realizations and Impulse/Response Sequences (2.1).

(i) We have seen in Part II (5.2) (iv) that in the continuous time-invariant case that the realization problem amounted to finding for a given sequence $\{A_i\}_{i \geq 1}$ of $p \times m$ matrices over k, fixed matrices (F,G,H) of appropriate sizes such that $A_i = HF^{i-1}G$ for all $i \geq 1$. The argument used δ-functions. Now in the discrete time case for $f: \Omega \to \Gamma$ we have an analogous situation. We define maps $\delta_i \in \Omega$ by $\delta_i(0) = e_i :=$ i-th unit basis vector of k^m (regarded as an $m \times 1$ column vector), and $\delta_i(t) = 0$ for $t \neq 0$. We let moreover $(f(\delta_i)_j)(t) :=$ j-th component of $(f(e_i))(t)$ where $(f(e_i))(t)$ is regarded as a column vector in k^p. Then from Section 1 it is easy to see that $\Sigma = (F,G,H)$ realizes f if and only if

$$(f(\delta_i)_j)(t) = (HF^{t-1}G)_{ji} :=$$ the element in the (j,i) position of the matrix $HF^{t-1}G$.

(ii) Thus again the realization problem in the discrete time case is equivalent to finding for a given sequence of $p \times m$ matrices $S = \{A_i\}_{i \geq 1}$ a fixed triple $\Sigma = (F,G,H)$ such that $A_i = HF^{i-1}G$ for all i. Σ is called a realization of S.

(iii) Given the discussion of Part II (5.2) a sequence $S = \{A_i\}_{i \geq 1}$ as in (i) is many times called an impulse/response sequence.

We now make the following classical definition:

Definition (2.2). Let $S = \{A_i\}_{i \geq 1}$ be any sequence of $p \times m$ matrices over k. Then the (i,j)-Hankel matrix (sometimes called the behavior matrix) for $i,j \in \mathbb{N}$ is defined to be

$$\mathcal{H}_{ij}(S) := \begin{bmatrix} A_1 & A_2 & \cdots & A_j \\ A_2 & A_3 & \cdots & A_{j+1} \\ \vdots & & & \\ A_i & A_{i+1} & \cdots & A_{i+j-1} \end{bmatrix}$$

We can now state the fundamental:

Theorem (2.3). Let $S = \{A_i\}_{i \geq 1}$ be as in (2.2). Then S is realizable if and only if there exist positive integers r,s such that $\operatorname{rank} \mathcal{H}_{rs}(S) = \operatorname{rank} \mathcal{H}_{r+1,s+j}(S)$ for all $j = 1,2,\ldots$. Moroever $n := \operatorname{rank} \mathcal{H}_{rs}$ is the dimension of a canonical realization.

Proof. Suppose that $S = \{A_i\}_{i \geq 1}$ is realizable by $\Sigma = (F,G,H)$. Set $R_j(F,G) := (G \; FG \; \ldots \; F^{j-1}G)$ and

$$Q_i(F,G) := \begin{pmatrix} H \\ HF \\ \vdots \\ HF^{i-1} \end{pmatrix} .$$

Then $Q_i(F,G)R_j(F,G) = \mathcal{H}_{ij}$ the (i,j)-Hankel of the squence $A_i = HF^{i-1}G$.

The proof of the converse is usually accomplished by constructing a specific realization for S given the rank conditions on the Hankel. We shall not give the technical details here but only remark that a key fact is that the block symmetry of the $\mathcal{H}_{ij}$ implies that the rank condition given above actually gives that $n = \text{rank}\ \mathcal{H}_{r,s} = \text{rank}\ \mathcal{H}_{r+i,s+j}$ for all $i,j \in \mathbb{N}$. The interested reader can find specific algorithms for constructing realizations in Ho [63], Ho and Kalman [64], and Silverman [133], [134] (as well as other places). A comparison of methods is given in [82], Chapter 10. See also the proof in (5.5) below.

Q.E.D.

Quotient Space of Canonical Systems (2.4).

(i) We make a bit of contact now with some of the partial realization theory of Section 4. Specifically, the result of (2.3) can be sharpened to show that for a given finite set (i.e. "a partial sequence") $A_1,\dots,A_{2n}$ of $p\times m$ matrices with the property that rank $\mathcal{H}_{n,n}$ = rank $\mathcal{H}_{n+1,n}$ = rank $\mathcal{H}_{n,n+1} = n$, there exists a canonical realization of dimension n unique up to the natural GL(n,k) action already discussed. In particular, given two canonical systems of dimension n , (F_1,G_1,H_1) and (F_2,G_2,H_2) with the same input and output spaces such that $H_1F_1^iG_1 = H_2F_2^iG_2$ for $i = 0,\dots,2n-1$, then there exists $g \in GL(n,k)$ such that $F_2 = g\,F_2\,g^{-1}$, $G_1 = gG_2$, $H_1 = H_2g^{-1}$. For details about this see [82], pages 301-303, and [54] (2.4.3).

(ii) In Part IV (we use the notation of Section 5 of Part IV) we discussed the moduli of completely reachable systems via the space $\mathcal{M}_{n,m,p}$. Recall that we were able to deduce the geometric properties of $\mathcal{M}_{n,m,p}$ almost formally using standard techniques from geometric invariant theory. In particular it was shown that $\mathcal{M}_{n,m,p}$ is quasi-projective. Similar remarks of course hold for the completely observable case.

(iii) This leaves us with canonical systems. Explicitly (we now take our base field k to be algebraically closed), let

$V^c_{n,m,p} := \{(F,G,H) \in M_{n,n}(k) \times M_{n,m}(k) \times M_{p,n}(k) \mid (F,G,H) \text{ is canonical}\}$.

Clearly $V^c_{n,m,p} \subset k^{n^2+nm+np}$ is a Zariski open subset which is invariant under the natural GL(n,k) action on $k^{n^2+nm+np}$ gotten by change of base in the state space k^n .

Note that $V^c_{n,m,p} \subset V_{n,m,p}$:= the space of completely reachable triples, and $V^c_{n,m,p}$ is also clearly a GL(n,k)-invariant open subset of this space. Since $(\mathcal{M}_{n,m,p},\pi_p)$ is a geometric quotient of $V_{n,m,p}$ (Part IV (5.3)), by definition (Part II (3.6)), there exists a Zariski open subset $\mathfrak{M}^c_{n,m,p}$ of $\mathfrak{M}_{n,m,p}$ such that $V^c_{n,m,p} = \pi_p^{-1}(\mathfrak{M}^c_{n,m,p})$. From what we have proven about $\mathfrak{M}_{n,m,p}$ it is immediate that $\mathfrak{M}^c_{n,m,p}$ is a smooth algebraic quasi-projective variety. We should remark here that even though we worked in Part IV, Section 5 over $\mathbb{C}$ for convenience, all the results there go easily through for an arbitrary algebraically closed field.

(iv) It is clear that $\mathfrak{M}^c_{n,m,p}$ should have certain natural moduli space properties. This will be discussed in Section 5 below.

(v) What is not so clear is the beautiful result of M. Hazewinkel [54] (2.5.7) that $\mathfrak{M}^c_{n,m,p}$ is not only quasi-projective but indeed quasi-affine. This result is the only result we know of in the theory of moduli of linear dynamical systems which does not come out almost immediately from Mumford's geometric invariant theory. In point of fact to prove it one needs realization theory, in particular Remark (2.4) (i) above.

We therefore want to prove now this important result:

Theorem (2.5). $\mathfrak{M}^c_{n,m,p}$ is quasi-affine.

Proof. First we define a morphism $\varphi\colon M_{n,n}(k) \times M_{n,m}(k) \times M_{p,n}(k) \to \mathbb{A}^{(n+1)^2mp}$ by

$$\varphi(F,G,H) := \begin{bmatrix} HG & HFG & \cdots & HF^nG \\ HFG & HF^2G & \cdots & HF^{n+1}G \\ \vdots & \vdots & & \vdots \\ HF^nG & HF^{n+1}G & \cdots & HF^{2n}G \end{bmatrix}$$

Notice that for $g \in GL(n,k)$, $\varphi(gFg^{-1}, gG, Hg^{-1}) = \varphi(F,G,H)$. Consequently if we restrict φ to $V^c_{n,m,p}$, φ descends to the quotient $\mathfrak{M}^c_{n,m,p}$ and defines a morphism $\tilde{\varphi}\colon \mathfrak{M}^c_{n,m,p} \longrightarrow \mathbb{A}^{(n+1)^2mp}$. But from (2.4)(i) we have then that $\tilde{\varphi}$ is injective.

Next denote the image of $\tilde{\varphi}$ by $\mathfrak{M}$. Now $\mathfrak{M}$ is a space of Hankel matrices of rank n, and consequently is clearly quasi-affine. Thus the morphism $\tilde{\varphi}\colon \mathfrak{M}^c_{n,m,p} \to \mathfrak{M}$ defines a bijection of $\mathfrak{M}^c_{n,m,p}$ with the quasi-affine variety $\mathfrak{M}$.

Unfortunately this is not enough to show that $\mathfrak{M}^c_{n,m,p}$ is quasi-affine since bijective morphisms of varieties need not be isomorphisms in the category of varieties (e.g. consider $\beta\colon \mathbb{A}^1 \to \mathbb{A}^2$ defined by $t \longmapsto (t^2, t^3)$). There are two methods we know of to show that $\tilde{\varphi}\colon \mathfrak{M}^c_{n,m,p} \longrightarrow \mathfrak{M}$ is in point of fact an isomorphism of varieties. The first method uses equations and inequalities defining the image $\tilde{\varphi}(\mathfrak{M}^c_{n,m,p}) \subset \mathbb{A}^{(n+1)^2mp}$ to write down local inverses (this is the method used by Hazewinkel [54] in the original proof of (2.5)). The second method is due to Byrnes [17] and while more elegant, makes use of some heavier machinery from algebraic geometry. Explicitly, one can show that $\mathfrak{M}$ is smooth (Clark [25]) and that $\tilde{\varphi}\colon \mathfrak{M}^c_{n,m,p} \longrightarrow \mathfrak{M}$ is birational (Falb [29]). Then from Zariski's Main Theorem

(Mumford [107]), the required result follows immediately.

Q.E.D.

§3. Realizations of Rational Matrices

In this section we discuss a very nice technique for the realization of rational matrices due to Fuhrmann [35]. This method has found many uses in the system theory literature (see [4], [36], and [37]) and is interesting because it combines the module theoretic approach of Kalman with the polynomial matrix methods of Rosenbrock [118]. Also since many of the ideas come from Hilbert space theory, Fuhrmann's realization and techniques could be the basis of a realization theory in the infinite dimensional case ([7], [38]). Finally we will need some of the facts from polynomial matrix theory and coprime factorizations in Part VIII.

We will let $M_{p,m}(k[z])$: = ring of $p \times m$ matrices with entries in $k[z]$, k an arbitrary field, and similarly we have the ring of rational matrices $M_{p,m}(k(z))$ where $k(z)$ is the quotient field of $k[z]$. Recall from Part II (5.1) that given a system $\Sigma = (F,G,H)$, $F \in M_{n,n}(k)$, $G \in M_{n,m}(k)$, $H \in M_{p,n}(k)$, the <u>associated transfer function</u> is $T_\Sigma(z) := H(zI-F)^{-1}G$. A system Σ <u>realizes</u> a given $T(z) \in M_{p,m}(k(z))$ if $T_\Sigma(z) = T(z)$. Finally $T(z) \in M_{p,m}(k(z))$ is <u>strictly proper</u> if the numerator of every element of $T(z)$ is of strictly lower degree than the denominator.

We first have:

<u>Proposition (3.1).</u> $T(z) \in M_{p,m}(k(z))$ is realizable if and only if it is strictly proper.

<u>Proof.</u> If $T(z)$ is realizable, then $T(z) = T_\Sigma(z)$ for some system Σ and $T_\Sigma(z)$ is patently strictly proper.

Conversely suppose that $T(z)$ is strictly proper. We may suppose without loss of generality that $p = m = 1$ i.e. that $T(z) \in k(z)$ since the entries of a $p \times m$ rational matrix consist of elements of $k(z)$, and if we can realize each entry individually, then we can realize the rational matrix by taking the direct sum of the realizations of the individual entries. Since $T(z)$ is strictly proper,

$$T(z) = \frac{c_{n-1} z^{n-1} + \dots + c_1 z + c_0}{z^n + d_{n-1} z^{n-1} + \dots + d_1 z + d_0} .$$

Set

$$F := \begin{bmatrix} 0 & 1 & 0 & \dots & 0 \\ 0 & 0 & 1 & \dots & 0 \\ \vdots & \vdots & \vdots & & \vdots \\ 0 & 0 & 0 & \dots & 1 \\ -d_0 & -d_1 & -d_2 & \dots & -d_{n-1} \end{bmatrix} , \quad g = \begin{bmatrix} 0 \\ \vdots \\ 0 \\ 1 \end{bmatrix}$$

$$h = [c_0 \; c_1 \; \dots \; c_{n-1}] .$$

Then it is easy to check $h(zI-F)^{-1}g = T(z)$.

Q.E.D.

Remarks (3.2).

(i) The realization of (3.1) is of course not in general not canonical. This is where Fuhrmann's theory comes in.

(ii) We let $z^{-1}k[[z^{-1}]]: =$ set of formal power series in z^{-1} with no constant term, and we have the corresponding matrices $M_{p,m}(z^{-1}k[[z^{-1}]])$. Note that an element $\ell \in M_{p,m}(z^{-1}k[[z^{-1}]])$ is of the form $\ell = \sum_{i \geq 1} A_i z^{-i}$ where the $A_i \in M_{p,m}(k)$. Hence associated to ℓ , we get a formal Hankel matrix $\mathcal{H}(\ell)$ defined in the obvious way. We set $Q_{p,m}: = \{\ell \in M_{p,m}(z^{-1}k[[z^{-1}]]) \mid \text{rank } \mathcal{H}(\ell) < \infty\}$.

Next we let $R_{p,m}: =$ {strictly proper elements of $M_{p,m}(k(z))$} which by (3.1) is = {transfer functions of systems with m inputs and p outputs} . Then the content of the realization Theorem (2.3) is that the following map $Q_{p,m} \to R_{p,m}$ is bijective: Let $\ell \in Q_{p,m}$, and canonically realize ℓ by (F,G,H) . Then we map $\ell \longmapsto H(zI-F)^{-1}G$.

We now set the background for constructing the Fuhrmann realization. For details about polynomial matrices see [35], [96] and [118].

Polynomial Matrices (3.3).

(i) An element $U \in M_{n,n}(k[z])$ is called unimodular if $\det U \in k - \{0\}$. It is well known ([96]) that $M_{n,n}(k[z])$ is a principal ideal ring (noncommutative for $n > 1$ of course) and in particular every pair of elements $F_1, F_2 \in M_{n,n}(k[z])$ have a greatest common right divisor and a greatest common left divisor unique up to unimodular factor. F_1, F_2 are right (resp. left) coprime if their greatest common right (resp. left) divisor is unimodular.

(ii) Given $f \in k(z)$, via the Euclidean algorithm we can write $f = s+r$ where $s \in k[z]$ and r is a strictly proper rational function. Let $k^p[z]$ denote the k-vector space of $p \times 1$ column vectors with coordinates in $k[z]$, and similarly for $k^p(z)$. Then we get a projection $\varphi: k^p(z) \to k^p(z)$ by sending

$$\begin{pmatrix} f_1 \\ \vdots \\ f_p \end{pmatrix} \longmapsto \begin{pmatrix} r_1 \\ \vdots \\ r_p \end{pmatrix}$$

where $r_i: =$ strictly proper part of f_i for $i = 1,\dots,n$.

(iii) Let $D \in M_{p,p}(k[z])$ with $\det D \neq 0$. Then we also get a projection $\pi_D: k^p[z] \to k^p[z]$ by defining $\pi_D(f): = D\varphi(D^{-1}f)$. It is immediate that $\ker \pi_D = D\, k^p[z]$. Now since $\det D \neq 0$, $k^p[z]/Dk^p[z]$ is a torsion $k[z]$-module and hence a finite dimensional vector space. Set $K_D: = \pi_D(k^p[z])$. Then K_D may be given a $k[z]$-module structure by setting $g \cdot f: = \pi_D(gf)$.

It is easy to see that $K_D \cong k^p[z]/Dk^p[z]$ as $k[z]$-modules.

(iv) Let $T \in M_{p,n}(k(z))$ be strictly proper. Then it is well-known (see [35]) that there exist $D \in M_{p,p}(k[z])$, $N \in M_{p,m}(k[z])$ with D invertible, D and N left coprime and unique up to common left unimodular factor such that $T = D^{-1}N$. This will be referred to as the left coprime factorization. One has an analogous existence and uniqueness result concerning a right coprime factorization $T = \tilde{N}\tilde{D}^{-1}$.

We can now construct:

Fuhrmann Realization (3.4). Fuhrmann's method actually gives two canonical realizations of a given strictly proper $T \in M_{p,m}(k(z))$ one corresponding to the left coprime factorization and the other corresponding to the right coprime factorization. So first let $T = D^{-1}N$ be a left coprime factorization as in (3.3)(iv). Now in general on $k^p[z]$ we have the shift operator σ: $k^p[z] \to k^p[z]$ defined by multiplication by z (compare with Kalman's construction in Section 1). Then if K_D is as in (3.3) (iii), we can restrict σ to K_D by setting $\sigma(D)(f) := \pi_D(\sigma f)$ for $f \in K_D$ and it is immediate that $\sigma(D): K_D \to K_D$ is a $k[z]$-module homomorphism.

Next regarding K_D as a k-vector space of dimension n, we have a linear map $G: k^m \to K_D$ defined by $Gv := N(z)v$. A key fact (see [35], page 533 (6.1)) is that the left coprimeness of D and N is equivalent to the fact that $(\sigma(D),G)$ is a completely reachable pair.

Finally we define a linear map $H: K_D \to k^p$ by $h(f) :=$ residue of $D^{-1}f$. Explicitly, since by definition of K_D, $D^{-1}f$ is strictly proper, we can expand $D^{-1}f = \sum_{j\geq 1} R_j z^{-j}$ where $R_j \in k^p$ for all $j \geq 1$. Then the residue of $D^{-1}f$ is defined as usual to be R_1.

Thus for our strictly proper rational function T, we claim we have a canonical realization $\Sigma = (\sigma(D),G,H)$ with K_D the state space. The fact that Σ is a realization is just a simple exercise of unwinding the definitions. We have already remarked that the coprimeness of D and N means that Σ is completely reachable. We will show that Σ is completely observable. So let $f \in K_D$ be such that $H\sigma(D)^i f = 0$ for all $i \geq 0$. Note however that

$$\begin{aligned} H\sigma(D)^i f &= \text{residue of } (D^{-1}\pi_D z^i f) \\ &= \text{residue of } (\varphi(D^{-1}z^i f)) \quad \text{(see (3.3)(ii) and (iii))} \\ &= R_{i+1} \end{aligned}$$

where $D^{-1}f = \sum_{j\geq 1} R_j z^{-j}$. Thus if $H\sigma(D)^i f = 0$ for all i we have that $D^{-1}f = 0$ and so $f = 0$.

Next let $T = \tilde{N}\tilde{D}^{-1}$ be a right coprime factorization. Then equating this factorization with the left coprime factorization above we get that $N\tilde{D} = D\tilde{N}$. It is not difficult to show (see [35], pages 429-431) that the map $A: K_{\tilde{D}} \to K_D$ given by $A(f) := \pi_D(Nf)$ is invertible. Moreover A is a homomorphism of $k[z]$-modules or equivalently A "intertwines" with the shifts $\sigma(D)$ and $\sigma(\tilde{D})$ i.e.

$A\circ\sigma(\tilde{D}) = \sigma(D)\circ A$. We can then construct linear maps $\tilde{G}: k^m \to K_{\tilde{D}}$, $\tilde{H}: K_{\tilde{D}} \to k^p$ with the defining properties $A\tilde{G} = G$, $HA = \tilde{H}$. Again it is easy to check that $(\sigma(\tilde{D}),\tilde{G},\tilde{H})$ is a canonical realization with state space $K_{\tilde{D}}$.

Finally the reader is urged to explicitly compare these constructions to those of Kalman described in (1.5) above.

§4. Partial Realizations

We have seen in Section 2, that the linear realization problem amounts to finding for a given infinite sequence of $p\times m$ matrices $\{A_i\}_{i\geq 1}$ a system $\Sigma = (F,G,H)$ of minimal dimension such that $HF^{i-1}G = A_i$ for all $i \geq 1$. Recall that this was based on the fact that the input/output map of a given system is completely determined by the sequence $HF^{i-1}G$.

Now physically it may be many times unrealistic to assume that one has total knowledge of the input/output map of a system i.e. the entire sequence $\{A_i\}_{i\geq 1}$. There can be many cases in which we have only part of the sequence and from this we must construct a model. This leads us to the partial realization problem which we would like to describe now. For simplicity we will assume $p = m = 1$ (the scalar input (output case) and that all our systems are defined over the real numbers $\mathbb{R}$. Many of the results generalize immediately.

Padé Approximation and Partial Hankel Matrices (4.1).

(i) The partial realization problem in its modern form has been formulated by Kalman in [77], [81] and [82]. We may state the problem as follows: Given a finite sequence $a_1,\ldots,a_r$ of real numbers, find all completions of the sequence to $a_1,\ldots,a_r,a_{r+1},\ldots$ such that $\sum_{i\geq 1} a_i z^{-i}$ is the Laurent expansion of a strictly proper rational function T with the degree of the denominator minimal. A canonical realization of T will be called a minimal partial realization of the sequence $a_1,\ldots,a_r$.

(ii) One also wants to parametrize all completions leading to minimal realizations of a given sequence. This is closely related to system identification problems since such realizations are perhaps the simplest models which account for the data $a_1,\ldots,a_r$. For more about this see our discussion in the introduction to Part VII as well as Section 4 of that part.

(iii) Actually the idea of partial realizations is equivalent to the classical idea of Padé approximation and as such goes back at least to Frobenius [33]. Namely, Padé approximation (which was not discovered by Padé) concerns the problem of finding for a given formal Laurent series $\sum_{i\geq 1} a_i z^{-i}$, a strictly proper rational function f with denominator of minimal degree whose Laurent expansion in z^{-1} has the property that the first r terms of the expansion agree with $a_1,\ldots,a_r$.

(iv) We would like to discuss now the solution to the partial realization problem given by Kalman in [81] since perhaps his is the neatest solution and moreover some of the invariants he introduces associated to formal power series will be of use to us in Part VII. So suppose we are given a finite sequence $a_1, \ldots, a_r$ of real numbers. Then to this sequence we may associate a partial Hankel matrix

$$\mathcal{H}_r(a) := \begin{bmatrix} a_1 & a_2 & a_3 & \cdots & a_r \\ a_2 & a_3 & a_4 & \cdots & * \\ \vdots & \vdots & \vdots & & \vdots \\ \vdots & a_r & * & \cdots & * \\ a_r & * & \cdots & \cdots & * \end{bmatrix}$$

Now the partial realization problem amounts to finding all completions (if they exist) such that the rank of $\mathcal{H}_r(a)$ remains constant where the rank is defined only modulo the elements $a_1, \ldots, a_r$; the unknown elements denoted by $*$ are ignored. Of course this is a simple problem in linear algebra which is solved by the following lemma (stated in [81] and a special case of which is in Hazewinkel [55]) whose proof we leave as an easy exercise:

Lemma (4.2). Given a matrix

$$\begin{bmatrix} A & b \\ c & * \end{bmatrix}$$

where $*$ is an undetermined element, then

(a) If both b and c are linearly dependent on A, then there is a unique value of $*$ which does not increase the rank of the matrix. Any other value will increase the rank by 1.

(b) If either b or c is linearly independent of A, then any value of $*$ will preserve the rank.

Partial Realizations (4.3). We would like now to explicitly indicate how (4.2) solves the problem of (4.1) and moreover gives some important invariants for power series. Let then $\{a_i\}_{i \geq 1}$ be an infinite sequence of real numbers. For each subsequence $a_1, \ldots, a_r$ associate the partial Hankel $\mathcal{H}_r(a)$. Then if $c_r :=$ number of columns of $\mathcal{H}_r(a)$ linearly independent modulo columns to the left, we have that $r - c_r$ = number of columns linearly dependent modulo columns to the left.

Suppose we adjoin to $a_1, \ldots, a_r$ the element a_{r+1}. Then we have

$$\mathcal{H}_{r+1}(a) = \begin{bmatrix} a_1 & \cdots & a_r & & a_{r+1} \\ a_2 & \cdots & a_r & a_{r+1} & * \\ \vdots & & & & \vdots \\ a_r & a_{r+1} & * & \cdots & * \\ a_{r+1} & * & * & \cdots & * \end{bmatrix}$$

(the triangle marked $\mathcal{H}_r(a)$ encloses the entries $a_1, \ldots, a_r$ in the upper left)

Now if an element a_{r+1} occurs in a row or a column linearly independent on the preceding ones, from (4.2) a_{r+1} does not change the rank of $\mathcal{H}_r(a)$. If an a_{r+1} occurs in a row and a column linearly independent on the preceding ones (this can happen $r+1-2c_r$ times when this quantity is positive from our above representation of $\mathcal{H}_{r+1}(a)$), if a_{r+1} does not have the unique value needed to preserve the rank of $_r(a)$, then the rank will "jump" by size $n_{r+1} = r+1-2c_r$. Such an $r+1$ is called a jump point of the sequence $\{a_i\}_{i\geq 1}$.

This gives us now a simple algorithm for solving the partial realization problem. Indeed let $a_1,\ldots,a_r$ be a given finite sequence which we want to minimally complete to $a_1,a_2,\ldots,a_r,a_{r+1},\ldots$ in the sense of (4.1) (i). This first jump point occurs at the minimal r_1 such that $a_{r_1} \neq 0$ and is of size $n_{r_1} = r_1$. Thus the dimension of a minimal partial realization of $a_1,\ldots,a_r$ is at least r_1. If we set $c_1 := c_{r_1}$ then for $r_1 \leq r \leq 2c_1$ we can have no jump points and hence the dimension of a minimal partial realization will remain r_1. Notice that in $a_1,\ldots,a_{r_1},\ldots,a_{2c_1}$ the parameters $a_{r_1+1},\ldots,a_{2c_1}$ are free in the sense that any choice of values for $a_{r_1+1},\ldots,a_{2c_1}$ will leave the dimension of a minimal partial realization of $a_1,\ldots,a_{r_1}$ invariant.

Now the next possible jump point is at some r_2 such that $r_2 > 2c_1$. If no such r_2 exists, then the dimension of a minimal realization is r_1. If such a r_2 does exist, then the dimension increases by size $n_{r_2} = r_2 - 2c_{r_2-1}$, and so on. Clearly then the dimension of a minimal realization is the sum of the sizes of the jump points. Always between two jump points the parameters are free.

Thus the above analysis answers the questions about the existence and parametrizations of minimal partial realizations. We will return to these questions from a more geometric point of view in Part VII, Section 4.

§5. Systems Over Rings

In Part IV we considered the notion of a family of systems defined over an algebraic variety. In this section we would like to concentrate on systems (and their realizations) which are defined over affine varieties or equivalently rings. More specifically we have:

Families of Systems (5.1).

(i) Recall from Part IV, Section 5, that a family of linear constant dynamical systems parametrized by a variety S consists of a vector bundle $V \to S$, a vector bundle endomorphism $\tilde{F}$, $(g_1,\ldots,g_m)$ global sections of $V \to S$, and $(h_1,\ldots,h_p)$ global sections of the dual bundle $V^* \to S$.

(ii) We suppose that S is an affine k-variety with coordinate ring $R := R_S$. We can consider V to be a locally free $\mathcal{O}_S$-module (see Part I, Section 5). Then it is well-known [52] that $X := V(S)$ (the global sections of V) is a finitely generated projective R-module. Moreover in the notation of Part I (5.5), regarding V as a locally free $\mathcal{O}_S$-module, $V \cong \tilde{X}$.

(iii) This clearly means that a family of systems in the sense of (i) over S affine is equivalent to the data (X,F,G,H) where X is a projective R-module, and $F: X \to X$, $G: R^m \to X$, $H: X \to R^p$ are R-module homomorphisms. Thus we get in a natural way an example of a system over the ring R.

(iv) For technical reasons we broaden the definition of a system over a ring to include X not necessarily projective. So we let R now be an arbitrary commutative ring with identity, and X be an arbitrary finitely generated R-module (X is called the state module). Then a system over the ring R, $\Sigma = (X,F,G,H)$ is a quadruple with F,G,H defined as in (iii).

Besides this algebro-geometric motivation for defining systems over rings, there are several physical motivations as well, some of which are detailed in Sontag's survey [137]. We would like to give here just one of these motivations:

Example (5.2). Consider the following delay-differential system over $\mathbb{R}$:

$$(*)\qquad \begin{aligned} \dot{x}_1(t) &= x_1(t-1) + x_2(t-\pi) + u(t) \\ \dot{x}_2(t) &= x_1(t-2) + 3x_2(t) + u(t-\pi) \\ y(t) &= x_1(t) + 2x_2(t-\pi) . \end{aligned}$$

In his very nice paper [83], Kamen found an algebraic approach for studying an infinite dimensional system of the type (*). (The system is infinite dimensional in the sense that the initial function needed to solve (*) belongs to a Banach state space of infinite dimension.)

Explicitly, following [83] we introduce two commuting delay operators $\sigma_1\varphi(t) := \varphi(t-1)$ and $\sigma_2\varphi(t) := \varphi(t-\pi)$ for any function $\varphi: \mathbb{R} \to \mathbb{R}$. Then (*) may be written as a system over the ring $\mathbb{R}[\sigma_1,\sigma_2]$ as follows:

$$\begin{pmatrix} \dot{x}_1 \\ \dot{x}_2 \end{pmatrix} = \begin{pmatrix} \sigma_1 & \sigma_2 \\ \sigma_1^2 & 3 \end{pmatrix} \begin{pmatrix} x_1 \\ x_2 \end{pmatrix} + \begin{pmatrix} 1 \\ \sigma_2 \end{pmatrix} u$$

$$y = (1 \quad 2\sigma_1) \begin{pmatrix} x_1 \\ x_2 \end{pmatrix} .$$

Note that since 1 and π are incommensurable the operators σ_1 and σ_2 are algebraically independent i.e. $\mathbb{R}[\sigma_1,\sigma_2]$ is isomorphic to a polynomial ring in two variables.

If one considers systems with commensurable delays, one gets systems defined over singular rings of course. (For example in (*) if we replace the delay 1 by $\sqrt{2}$ and the delay π by $\sqrt[3]{2}$, we will get a system defined over $\mathbb{R}[\sigma_1,\sigma_2]/(\sigma_1^2 - \sigma_2^3)$.)

Remark (5.3). Having motivated the study of systems over rings we now want to study the corresponding realization theory. The basic reference for this is Eilenberg [28]. We first make the following obvious generalizations of some of the preceding definitions:

Definitions (5.4).

(i) Let $\Sigma = (X,F,G,H)$ be a system over R. Assume that X may be generated by n elements over R. Then Σ is completely reachable if the homomorphism $R_n : R^{nm} \to X$ defined by $(G\ F\circ G \ldots F^{n-1}\circ G)$ is surjective. Σ is completely observable if the homomorphism $Q_n : X \to R^{np}$ defined by

$$\begin{pmatrix} H \\ H\circ F \\ \vdots \\ H\circ F^{n-1} \end{pmatrix}$$

is injective. Σ is canonical if it is completely reachable and completely observable.

(ii) Let $S = \{A_i\}_{i\geq 1}$ be a sequence of $p\times m$ matrices over R. One can define the associated Hankel matrices exactly as in (2.2). It will also be convenient to define a formal matrix, the Hankel matrix of S, by

$$\mathcal{H}(S) = \begin{bmatrix} A_1 & A_2 & \cdots\cdots \\ A_2 & A_3 & \cdots\cdots \\ \vdots & \vdots & \\ \vdots & \vdots & \end{bmatrix}$$

which contains the $\mathcal{H}_{ij}(S)$ as submatrices. $\mathcal{H}(S)$ has finite rank n if for some r,s $n = \operatorname{rank} \mathcal{H}_{r,s}(S) = \operatorname{rank} \mathcal{H}_{r+i,s+j}(S)$ for all non-negative integers i,j. Then from (2.3) (and its proof) we see that over a field k, S is realizable if and only if $\mathcal{H}(S)$ has finite rank.

(iii) More generally given $S = \{A_i\}_{i\geq 1}$ over a ring R, a realization of S is a system Σ such that $A_i = H\circ F^{i-1}\circ G$ for all $i \geq 1$.

(iv) Notice that if $S = \{A_i\}_{i\geq 1}$ admits a realization $\Sigma = (X,F,G,H)$, then it admits a realization $\tilde{\Sigma} = (\tilde{X},\tilde{F},\tilde{G},\tilde{H})$ where $\tilde{X}$ is a free R-module. Indeed suppose that X is generated by n elements (by definition we always take our state modules to be finitely generated). Then there exists a surjective homomorphism $R^n \to X$. Hence we can always find $\tilde{F}$, $\tilde{G}$, $\tilde{H}$ such that the following diagram is commutative:

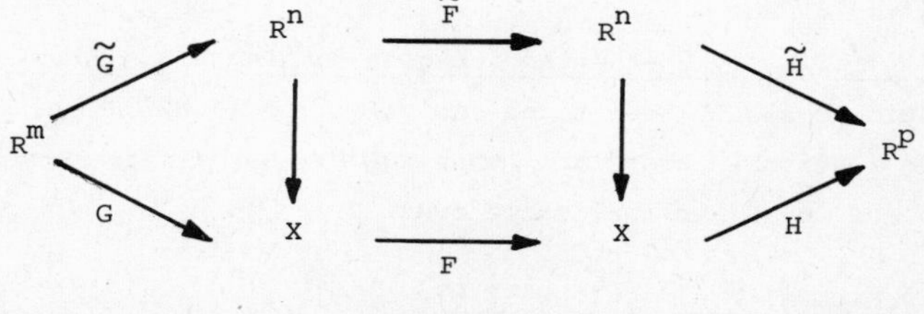

Thus if S is realized by Σ it is also realized by $(R^n,\tilde{F},\tilde{G},\tilde{H})$.

We would like to give now Sontag's neat proof [137], pages 28-29 to one of the key results in realization theory over rings.

Theorem (5.5). A sequence of $p\times m$ matrices over a Noetherian integral domain R , $S = \{A_i\}_{i\geq 1}$ is realizable if and only if rank $\mathcal{H}(S)$ is finite.

Proof. Clearly if S is realizable, then rank $\mathcal{H}(S)$ is finite (see the proof of (2.3)). For the converse, we let X be the R-module generated by the columns of $\mathcal{H}(S)$. Since $\mathcal{H}(S)$ by hypothesis has finite rank, say n , we can find columns $c_1,\ldots,c_n$ which generate the K-vector space generated by the columns of $\mathcal{H}(S)$ where $K:=$ quotient field of R . But by Cramer's rule given any column c , there exist $a_1,\ldots,a_n$, $a \in R$ such that $c = \Sigma(a_i/a)c_i$. Thus X is a submodule of the R-module generated by the c_i/a , which means that X is finitely generated since R is Noetherian.

To complete the proof we realize (this realization was already used e.g. by Rouchaleau [119] and Fliess [31]) S by a canonical system with state module X . We let b_i be the i-th column of $\mathcal{H}(S)$. Then $F: X \to X$ is defined by extending linearly the shift operator which sends any column b_i to b_{i+m} . This is well-defined because of the block symmetry of $\mathcal{H}(S)$ (recall each A_j is $p\times m$). $G: R^m \to X$ is defined by $G(a_1,\ldots,a_m) = a_1b_1 + \ldots + a_mb_m$, and $H: X \to R^p$ is defined by taking the image of any column of $\mathcal{H}(S)$ the vector made up of the first p elements of that column. It is then easy to see that $\Sigma = (X,F,G,H)$ is a canonical realization of $\mathcal{H}(S)$ proving the theorem.

Q.E.D.

Definitions-Remarks (5.6).

(i) Let R be an integral domain with quotient field K . Then R is said to be Fatou if any sequence $S = \{A_i\}_{i\geq 1}$ of $p\times m$ matrices over R which is realizable over K is realizable over R . From (5.4) (iv) we can always assume that the state module of the realization is free. So in other words, the property of Fatou means that if for given S we can find matrices over K , (F,G,H) such that $A_i = HF^{i-1}G$ for all i , then there exist $(\tilde{F},\tilde{G},\tilde{H})$ matrices over R such that $A_i = \tilde{H}\tilde{F}^{i-1}\tilde{G}$ for all i .

(ii) This terminology comes from the fact that Fatou [30] showed that $\mathbb{Z}$ is a Fatou ring.

We thus have the following theorem due to Rouchaleau-Wyman-Kalman [122] (see also Eilenberg [28], page 439) :

Theorem (5.7). Every Noetherian integral domain R is Fatou.

Proof. This is easily seen to be equivalent to (5.6). For example if $S = \{A_i\}_{i\geq 1}$ is realizable over K , then $\mathcal{H}(S)$ has finite K-rank and therefore finite R-rank, and so S is realizable over R by (5.5).

Q.E.D.

Absolutely Minimal Realizations (5.8).

(i) We have seen in (1.7) that over a field a canonical realization $S = \{A_i\}_{i\geq 1}$

is also minimal. Now we have also seen that over a Noetherian integral domain R if we can realize S over the quotient field K, then we can realize S over R, but the realization may of course have larger dimension.

(ii) In [120], Rouchaleau and Sontag study the question of when a given S has a realization over R the same dimension as a minimal realization of S over the quotient field K. They call such realizations absolutely minimal.

(iii) Now while it is true that every S has a canonical realization over R unique up to isomorphism (see Eilenberg [28] Chapter 16, Section 5), it is not true that every S admits an absolutely minimal realization. In point of fact, in [120] (see (4.2), (5.1), and (5.3)) the following results are proven:

(a) The canonical realization of every sequence $S = \{A_i\}_{i \geq 1}$ over R a Noetherian domain is absolutely minimal if and only if R is a principal ideal domain.

(b) For R a polynomial ring over a field, every realizable S has an absolutely minimal realization if and only if the ring R has no more than two variables.

(iv) We have mentioned these results because of their theoretical interest and also to indicate how many times results about realizations over fields need not always generalize to the case of an arbitrary ring. This situation also occurs for a notion of complete observability which we are about to define, which coincides with the old notion over a field, but which is a stronger condition over an arbitrary Noetherian ring. This notion which is due to Sontag [138] is motivated for example in the construction of observers for delay differential systems.

The precise definitions that we will need are:

Definitions (5.9).

(i) For R a commutative ring with identity, and M an R-module we set $M^* := \mathrm{Hom}_R(M,R)$ the dual module of M. Note that given M, N R-modules and $f: M \to N$ an R-homomorphism, we get an R-homomorphism $f^t: N^* \to M^*$ ($f^t :=$ the transpose of f) by $f^t(\varphi) := \varphi \circ f$ for $\varphi \in N^*$.

(ii) Given a system $\Sigma = (X,F,G,H)$ over R, the dual system $\Sigma^* := (X^*,F^t,H^t,G^t)$.

(iii) A split system $\Sigma = (X,F,G,H)$ over R, is a system such that X is a finitely generated projective R-module with R a Noetherian ring, and such that Σ and Σ^* are both completely reachable.

Split and Canonical Systems (5.10).

(i) It is clear that if Σ is split, then Σ is canonical. Over a field the converse is also true. However over an arbitrary ring this may not be the case. For example, over $\mathbb{Z}$ consider the system $(\mathbb{Z}, 1, 1, 2)$ (i.e. input dimension = output dimension = 1). It is immediate that this system is canonical. However the dual system is $(\mathbb{Z}, 1, 2, 1)$, which is not completely reachable, and so $(\mathbb{Z}, 1, 1, 2)$ is not split. In this sense then, we get a stronger notion of observability.

(ii) We have been using tacitly the fact that complete reachability is a local property. More precisely, given R a Noetherian ring, $m \subset R$ a maximal ideal, we let $k(m) := $ the residue field R/m. Given a system Σ defined over R, in the obvious way, we get a system $\Sigma(m)$ defined over $k(m)$ for each maximal $m \subset R$. From elementary commutative algebra (e.g. [6]), it is easy to show that Σ is completely reachable if and only if $\Sigma(m)$ is completely reachable for every maximal $m \subset R$. This justifies our taking in the definition of completely reachable family over a variety S (see Part IV, Section 3), the complete reachability defined locally in each fiber.

(iii) However, complete observability is not a local property (e.g. for $\Sigma = (\mathbb{Z}, 1, 1, 2)$ consider the corresponding localization at the maximal ideal $2\mathbb{Z}$). Thus we have a problem of defining (via the recipe of Part IV) a "correct" and system theoretically natural functor for the quotient space of canonical systems $\mathcal{M}^c_{n,m,p}$ (notation as in (2.4)) to represent. We will see below this is naturally solved through the notion of "split system" giving a moduli-theoretic motivation for introducing this definition.

(iv) We are going to prove now the fundamental result concerning split systems, a result first proven by Sontag [138], page 32 and then by Byrnes [17], page 1349. Even though this result is true over more general rings, for our purposes it will be enough to take the ring R to be a finitely generated k-algebra (k an algebraically closed field) such that R cannot be written as a product of k-algebras $R_1 \times R_2$. This latter condition is equivalent to the variety associated to R being connected as a topological space [6]. For general finitely generated k-algebras, the result (5.11) below can be generalized by just assuming that the Hankel (see that statement of (5.11)) is of constant rank on each connected component of the corresponding variety.

Theorem (5.11). Let R be as in (5.10) (iv) and let $\Sigma = (X,F,G,H)$ be a canonical system over R with associated Hankel matrix $\mathcal{H}$ (i.e., $\mathcal{H}$ is the Hankel matrix associated to the sequence $\{H \circ F^{i-1} \circ G\}_{i \geq 1}$). Then Σ is split with X of rank n if and only if $\mathrm{rank}_k \mathcal{H}(m) = n$ for all maximal ideals $m \subset R$.

Proof. First suppose that Σ is split with X of rank n. Then it is easy to see from the case of a field, that to show rank $\mathcal{H}(m) = n$ for all maximal ideals m, it suffices to show that $\Sigma(m)$ is canonical for every m. (Explicitly, this is because by the projectivity of X and the connectedness of R, $n = \dim_k X/mX = \mathrm{rank}_R X$. Then $\Sigma(m)$ being canonical implies that $\mathrm{rank}_k \mathcal{H}(m) = n$.) Now since complete reachability is a local property, $\Sigma(m)$ is completely reachable and $\Sigma^*(m)$ is completely reachable for $m \subset R$ maximal, where Σ^* is the dual system. However from the projectivity of X,

$\Sigma^*(m) \cong \Sigma(m)^*$. But this implies that $\Sigma(m)$ is completely observable.

Conversely suppose we have $\operatorname{rank}_k \mathcal{H}(m) = n$ for all $m \subset R$ maximal. Suppose we have proven that X is projective of rank n . Then it is almost trivial to prove that Σ is split. Indeed under the above supposition we have $\operatorname{rank}_k X/mX = n$ for all m , and thus by (2.3) and the fact that over a field the concepts of minimal and canonical realization coincide, we have that $\Sigma(m)$ is canonical for each maximal $m \subset R$. Thus Σ is completely reachable. But as before since X is projective $\Sigma(m)^* \cong \Sigma^*(m)$ and thus $\Sigma^*(m)$ is completely reachable for all m , i.e. Σ^* is completely reachable, proving that Σ is split.

We show now that the hypothesis on the Hankel implies that X is projective of rank n . Recall from Part I, Section 3 that to R we may associate an affine variety V_R in such a way that the maximal ideals of R correspond to the points of V_R (this is the Hilbert Nullstellensatz). We assume that the space of input values of Σ is R^m , and the space of output values is R^p . Then the fact that $\operatorname{rank}_k \mathcal{H}(m) = n$ for all $m \subset R$ maximal, implies that if we let $\mathcal{M}$ be the quasi-affine space of Hankels of rank n over k with block matrix entries of size $p \times m$, then we get a morphism from $V_R \to \mathcal{M}$. But from the proof of (2.5) $\mathcal{M}^c_{n,m,p} \cong \mathcal{M}$ and hence we get a morphism from $V_R \to \mathcal{M}^c_{n,m,p}$ which in turn determines a morphism $f\colon V_R \to \mathcal{M}_{n,m,p}$. Using the notation of Part IV (5.4) we have that $\tilde{\mathcal{F}}_p(V_R) \cong \operatorname{Hom}(V_R, \mathcal{M}_{n,m,p})$ i.e. associated to f we get a completely reachable family unique up to isomorphism of rank n parametrized by V_R . But from Part I, Section 5 the vector bundle of this family regarded as a locally free $\mathcal{O}_{V_R}$-module is of the form $\tilde{X}_1$ where X_1 is a projective R-module of rank n (see the notation in Part I (5.5)) and similarly we derive a triple of R-homomorphisms $F_1\colon X_1 \to X_1$, $G_1\colon R^m \to X_1$, $H_1\colon X_1 \to R^p$. By construction it is also clear that the system $\Sigma_1 = (X_1, F_1, G_1, H_1)$ is canonical and has the Hankel matrix $\mathcal{H}$, i.e. Σ_1 is a canonical realization of the sequence $\{H \circ F^{i-1} \circ G\}_{i \geq 1}$. But by the uniqueness of canonical realizations ([28] Chapter 16, Section 5) we have that $\Sigma \cong \Sigma_1$ and thus X is projective.

Q.E.D.

<u>Corollary (5.12).</u> Let R be as in (5.11). Then a canonical system Σ over R is split if and only if $\Sigma(m)$ is observable for all maximal $m \subset R$.

<u>Proof.</u> A complete proof of this important result due to C. Byrnes may be found in [17], pages 1349-1351. We would just like to remark here that besides (5.11) the other key ingredient in the proof is a generalization of a bundle defined by Hermann-Martin [98] which will come up naturally in the dimension of state feedback of Part VIII, Section 2.

Q.E.D.

<u>Split Families (5.13).</u>

(i) Over an arbitrary k-variety S we say that a family of systems Σ in $\tilde{\mathcal{F}}_p(S)$ (notation as in Part IV (5.4)) is <u>split</u> if the corresponding system in the fiber over $s \in S$, $\Sigma(s)$, is canonical for all $s \in S$. By (5.12) this agrees with (5.5) (iii) for affine varieties.

(ii) If we denote by $\tilde{\mathfrak{F}}_p^s$ the subfunctor of $\tilde{\mathfrak{F}}_p$ consisting of families of split systems, then trivially $\mathfrak{M}^c_{n,m,p}$ represents $\tilde{\mathfrak{F}}_p^s$.

§6. Polynomial Systems

So far we have been essentially concerned with linear systems defined over a ring or a field. Now over a field k, there is a more general class of systems which one would naturally attempt to apply the methods of algebraic geometry, namely discrete systems whose input/output maps are polynomial. These systems can be realized by appropriate combinations of multipliers, adders, delay lines and amplifiers and hence carry great generality and practical importance.

Multilinear input/output maps were already studied in the late 1960's by Kalman [76]. Polynomial systems from an algebro-geometric viewpoint have been studied by Sontag-Rouchaleau [141], and Sontag [139], [140]. In this section we would like to report on some of this work. The full theory is too long and technical to describe here in any detail and so we leave it to the interested reader to look up the details in the above references. We might add that the results here are completely independent of those given in the sequel and so those interested just in the linear theory can skip this section.

One of the interesting features of polynomial systems is that in order to get a satisfactory theory of minimal realizations for polynomial input/output maps (to be defined explicitly below), the category of k-varieties is too small, and so one must go to a larger class of objects, namely k-schemes of not necessarily finite type. More specifically we have:

Affine k-Schemes (6.1). In Part I, Sections 1 and 3 we discussed the basic concepts of affine algebraic geometry over an algebraically closed field $\bar{k}$, and then in Section 7 we sketched how these ideas may be extended to arbitrary fields. Recall in particular that associated to an affine variety X is the coordinate ring R_X which we showed could be identified with the ring of regular functions on X.

Now R_X is a finitely generated reduced $\bar{k}$-algebra. (By reduced we mean that R_X has no nilpotent elements or equivalently [6] the intersection of all its prime ideals i.e. its nilradical is (0). Note moreover in R_X the nilradical is equal to the Jacobson radical which is defined to be the intersection of all the maximal ideals. For proofs of all this see Chapter 1 of [6].) Moreover via the Hilbert Nullstellensatz we identified the points of X with the $\bar{k}$-algebra homomorphism $R_X \to \bar{k}$. Finally given a finitely generated $\bar{k}$-algebra R, we can write $R = k[X_1,\ldots,X_n]/I$, and I will define a $\bar{k}$-variety in $\mathbb{A}^n$ with coordinate ring $\bar{R} = k[X_1,\ldots,X_n]/\sqrt{I}$. In particular, from the point of view of constructing varieties for $\bar{k}$-algebras we may always assume that the $\bar{k}$-algebras are reduced.

Now these results go over essentially unchanged in case k is an arbitrary infinite field. Indeed given R an arbitrary k-algebra, we define a k-ideal to be the kernel of a k-algebra homomorphism $R \to k$, and the radical is taken to be the intersection of all k-ideals. R is reduced if and only if the radical is (0). It is easy to show that R is reduced if and only if it is isomorphic to a k-algebra of functions i.e. a subalgebra of the algebra (under pointwise operations) defined by the set of all functions $X \to k$ for some set X. (For a proof of this as well as the other remarks made below see Bourbaki [11], Dieudonne [27], and Sontag [139].)

We can now play the familiar game of Part I. Indeed given a reduced k-algebra R, set $V_R := \mathrm{Hom}_k(R,k)$. For any subset $S \subset R$, we define $V(S) := \{p \in V_R \mid p(a) = 0 \text{ for all } a \in S\}$ and as before one can show that the $V(S)$ are the closed sets for a topology on V_R called the Zariski topology. Given a closed set one can as in Part I, Section 1 define the associated ideal and one has a result analogous to that of Theorem (1.11) of that part.

Consequently one can define an affine k-scheme to be a pair (X,R_X) consisting of a set X and a k-algebra R_X of functions $X \to k$ such that the map $X \to \mathrm{Hom}_k(R_X,k)$ defined by $p \longmapsto \varphi_p$ where $\varphi_p \in \mathrm{Hom}_k(R_X,k)$ is $\varphi_p(a) := a(p)$ for each $a \in R_X$, is a bijection. The elements of R_X are called (of course) regular functions. The bijection $X \xrightarrow{\sim} \mathrm{Hom}_k(R_X,k)$ induces a topology on X via the Zariski topology. A morphism of affine k-schemes $(X,R_X) \to (Y,R_Y)$ is a continuous map $\varphi: X \to Y$ such that for every $f \in R_Y$, $f \circ \varphi \in R_X$.

Finally note that for $k = \bar{k}$, R_X finitely generated, this definition of affine k-scheme is precisely the classical construction given above for constructing an affine variety from a reduced k-algebra.

We can now make the following definition:

Definition (6.2). A polynomial system Σ over an arbitrary infinite field k is a discrete constant dynamical system (in the sense of (1.5) and (1.9) of Part II) such that

(i) X,U,Y are affine k-schemes (X = state space, U = set of input values, Y = set of output values).

(ii) The state transition map $\varphi: X \times U \to X$ and the readout map $\eta: X \to Y$ are morphisms.

(iii) There is given an equilibrium state $x_0 \in X$, i.e. a state such that there exists $u_0 \in U$ with $\varphi(x_0,u_0) = x_0$.

Remarks (6.3).

(i) Our use of the term "polynomial system" is broader than that used in [139], [141]. In [139] "polynomial system" refers to the case in which X, U, Y are varieties i.e. such that the associated k-algebra R_X, R_U, R_Y are finitely generated and U is irreducible (see (2.1), page 58 of [139]). What we have called "polynomial system" in (6.2) would be referred to as "k-system" in [139] ((8.1), page 60). However, for our sketch here we will not need two terms, and so we will use "polynomial system" in this larger context.

(ii) Note that in case $X \cong k^n$, $U \cong k^m$, $Y \cong k^p$, Σ is defined by polynomial difference equations of the form

$$x(t+1) = \varphi(x(t),u(t))$$
$$y(t) = \eta(x(t))$$

from which the term "polynomial system" comes.

(iii) Even though many of the results below are true under more general conditions for our sketch here we will assume for now on that $U \cong k^m$, $Y \cong k^p$, and X is an arbitrary affine k-scheme. Moreover we assume that the equilibrium point $x_0 \in X$ is such that $\varphi(x_0,0) = x_0$.

We now want to define a proper notion of "reachability" in the polynomial context:

Quasi-Reachability (6.4).

(i) If U^t is the Cartesian product of U taken t times for $t \geq 0$ (where $U^\circ := \{\emptyset\}$ consists of a unique element), then we set $U^* :=$ disjoint union of the U^t for all $t \geq 0$. On U^* define an equivalence relation $\sim$ by $(u_1,\ldots,u_t) \sim (u_1,\ldots,u_t,0,\ldots,0)$. Then $U^*/\sim$ may be regarded as the set of all infinite sequences with finitely many non-zero entries. If we introduce an indeterminate z, then clearly $U^*/\sim$ is isomorphic to $u[z] := \{u_t z^{t-1} + \ldots + u_1\}$ where $(u_1,\ldots,u_t,0,\ldots)$ corresponds to $u_t z^{t-1} + \ldots + u_1$.

(ii) Given our transition map $\varphi: X \times U \to X$ we can inductively extend φ to sequences of inputs as follows:

$\varphi^{(t+1)}(x,u_1,\ldots,u_{t+1}) := \varphi(\varphi^{(t)}(x,u_1,\ldots,u_t),u_{t+1})$.

Then we get a map $\varphi: X \times U[z] \to X$ defined by $\tilde{\varphi}(x,u) := \varphi^{(t)}(x,u)$ for $u \in U^t$. By our assumptions, this is well-defined. Note that in this definition, we are tacitly identifying U^t with its image in $U[z]$ which we also denote by U^t. In particular, $\tilde{\varphi}|U^t = \varphi^{(t)}$ and the restrictions of $\tilde{\varphi}$ to the U^t are morphisms.

(iii) We come now to the crucial definitions. First define the <u>reachability map</u> $g: U[z] \to X$ by $g(u) = \tilde{\varphi}(x_0,u)$. Again $g|U^t$ is a morphism of k-schemes. Then Σ is called <u>reachable</u> if g is surjective. The physical motivation for this should be clear from Part II, Section 2. Σ is called <u>quasi-reachable</u> if $\overline{\bigcup_{t\geq 0} g(U^t)} = X$, the closure being taken in the Zariski topology.

(iv) The <u>dimension</u> of Σ will be defined to be the <u>dimension</u> of the state space X as a scheme which in turn is defined in analogy with the classical case (see Part I, Section 3) to be the transcendence degree of the associated k-algebra R_X over k. Note that if R_X is not an integral domain then $\text{trans.deg.}R_X := \sup_{\substack{P\subset R_X \\ P \text{ prime}}} \text{trans. deg.}R_X/P$. Of course this dimension need not be finite.

We now have the following result from [141], page 59:

Proposition (6.5). If $\dim \Sigma = n$, then

$$\overline{g(U^n)} = \overline{\bigcup_{\tau \geq 0} g(U^t)} \quad \text{(notation as in (6.4)).}$$

Proof. It is easy to show that the sequence of sets $\overline{g(U^t)}$ is a monotonically increasing sequence such that if for some t_0, $\overline{g(U^{t_0})} = \overline{g(U^{t_0+1})}$, then we have equality from the point t_0 on. Now for each irreducible $\overline{g(U^t)}$ we have the corresponding prime ideal I_t, and the set of prime ideals $\{I_t\}_t$ is monotonically decreasing, and if we have equality $I_{t_0} = I_{t_0+1}$, then we have equality from the point t_0 on. But then by standard dimension theory (e.g. [6], Chapter 11) this means that the sequence of ideals must be such that $I_n = I_{n+1} = I_{n+2} = \ldots$ which proves the proposition.

Q.E.D.

Remarks (6.6).

(i) For finite dimensional polynomial systems this means a system is quasi-reachable if and only if it is quasi-reachable in bounded time. The analogous property does not hold for reachability ([141], page 60) which is one reason why the concept of quasi-reachability is easier to work with. More importantly if a system is quasi-reachable, then by definition the reachable states are dense (the set of reachable states being $\bigcup_{t \geq 0} g(U^t)$) and hence the dynamics are in point of fact already determined on a Zariski dense subset. Thus quasi-reachability seems to be the more natural concept for polynomial systems.

(ii) We now will need an appropriate definition for "observability". We have seen in Part II, Section 3 that observability means that one can gain full knowledge of the states of the system from processing the input/output data. In the polynomial case, it seems natural that the definition should be strengthened to require that this knowledge should be gotten from a polynomial processing of the input/output data. For an extensive discussion of such questions see [140]. For our purposes however the following will suffice:

Algebraic Observability (6.7).

(i) Given $u \in U^t$, define a morphism $\eta^u: X \to Y$ by $\eta^u(x) := \eta \circ \varphi^{(t)}(x,u)$. (We are using the notation of (6.2) and (6.4)). If we set $\Gamma := \{\text{maps } U^* \to Y\}$, then we have a map $\mu: X \to \Gamma$ defined by $\mu(x) :=$ the map $U^* \to Y$ which sends $u \longmapsto \eta^u(x)$ for $u \in U^*$. Σ is said to be observable if μ is injective.

(ii) For each $j = 1,\ldots,p$, let $\pi_j: Y \to k$ be a projection on the j-th factor ($Y \cong k^p$). The observables of Σ are the regular functions $\eta_j^u := \pi_j \circ \eta^u: X \to k$. We define the observation algebra A_Σ to be the k-subalgebra of R_X generated by the observables. Σ is algebraically observable if and only if $A_\Sigma = R_X$.

(iii) Algebraic observability is a stronger property than observability. Indeed to see this note that $\Gamma = Y^{U^*}$ the direct product of Y with itself taken U^* times (i.e. $\prod_{w \in U^*} Y$) and thus may be given a natural structure as a k-scheme (see [27]). Now A_Σ being a subalgebra of R_X is reduced, and hence defines the affine k-scheme $V: = \mathrm{Hom}_k(A_\Sigma, k)$ whose algebra of regular functions is A_Σ. Then the inclusion $A_\Sigma \hookrightarrow R_X$ induces a morphism $X \to V$. Moreover $V \hookrightarrow \Gamma$ as a closed subscheme and the composition $X \to V \hookrightarrow \Gamma$ is nothing but μ. (To see this explicitly one can note that R_Γ is generated by the algebras R_Y appearing in the coproduct indexed by U^* i.e. the dual of the discrete product Y^{U^*} is such a coproduct. From this it is easy to show that the homomorphism associated to μ by duality $\mu^*: R_\Gamma \to R_X$ has image A_Σ.) Then to say Σ is algebraically observable means that $X = V$ and hence μ is a closed immersion of schemes and not merely injective as in the observable case.

(iv) A polynomial system Σ is <u>algebraically canonical</u> if it is quasi-reachable and algebraically observable. It is these systems which will play the fundamental role in the realization theory of polynomial input/output maps which we are about to describe.

<u>Realizations of Polynomial Input/Output Maps (6.8).</u> We are now ready to describe some polynomial realization theory. The reader should notice that many of the ideas given here are straightforward generalizations of the ideas described in Section 1 for the linear case.

Accordingly we first need a notion of "input/output map" which takes input sequences to output sequences. We set

$$U_s: = \{u: \mathbb{Z} \to U \mid \text{there exists } t_u \text{ such that } u(t) = 0 \text{ for all } t < t_u\} ,$$
$$Y_s: = \{y: \mathbb{Z} \to Y \mid \text{there exists } t_y \text{ such that } y(t) = 0 \text{ for all } t < t_y\} .$$

Then a <u>constant causal input/output map</u> is a map $f: U_s \to Y_s$ such that

(a) for every $u_1, u_2 \in U_s$ and $\tau \in \mathbb{Z}$, if for $t < \tau$ we have $u_1(t) = u_2(t)$ then $f(u_1)(t) = f(u_2)(t)$ for $t \leqslant \tau$ (i.e. the output depends only on values of the input $u(t)$ for $t < \tau$; this is called <u>causality</u>);

(b) if $\sigma_{U_s}: U_s \to U_s$ is defined by $\sigma_{U_s}(t): = \sigma_{U_s}(t+1)$ and similarly for $\sigma_{V_s}: V_s \to V_s$, then $f(\sigma_{U_s}(u)) = \sigma_{V_s} f(u)$ for every $u \in U_s$ (this is called <u>constancy</u>).

Now we shall need a notion of <u>polynomial input/output map.</u> To define this we will need the concept of <u>Volterra series</u>. By definition this is formal power series over k in countably many variables which is of finite degree in each variable separately. Suppose temporarily that $p = m = 1$ i.e. our $f: U_s \to Y_s$ is a constant causal scalar input/output map. Such a map is clearly uniquely determined by defining how $y(1)$ depends on $u(t)$ for $t \leqslant 0$. Then provisionally we will say that f is polynomial if there exists a Volterra series α in the variables $z_1, z_2, \ldots, z_n, \ldots$ such that $y(1)$ which depends on an input sequence $u \in U_s$ is gotten by substituting $u(1-t)$ for z_t in α. This makes sense from the definition of U_s, and from the fact that given any finite subset of $\{z_i\}_{i\geq 1}$, α will be a polynomial in the variables of this finite subset. For example the standard linear system with impulse response sequence $\{a_i\}_{i\geq 1}$ ((2.1)(iii)) may be represented by the Volterra series $a_1z_1 + a_2z_2 + \ldots + a_nz_n + \ldots$.

More generally (and formally) we will let

$Q := \{$Volterra series over k in the countably many variables

z_{ij}, $i = 1,\ldots,m$, $j = 1,2,3,\ldots\}$.

(The reason for the double index will be made clear shortly.) Q may be given a k-algebra structure in the obvious way and it is easy to see that Q is reduced. (For complete details about Volterra series see [139], pages 42-50.)

In the standard way, set $z_j := (z_{ij}, \ldots, z_{mj})$ for each j so that formally we may regard $\alpha \in Q$ as a power series in $z_1, z_2, \ldots$. Then we get a surjective homomorphism of k-algebras $r_t: Q \to k[z_1,\ldots,z_t]$ by sending $\alpha(z_1,z_2,\ldots,z_t,z_{t+1},\ldots) \longmapsto \alpha(z_1,z_2,\ldots,z_t,0,0,\ldots)$. But notice that $k[z_1,\ldots,z_t]$ is isomorphic to the algebra of regular functions on U^t i.e. $R_{U^t} \cong k[z_1,\ldots,z_t]$. (Notice that the regular function z_{ij} acts on U^t by $z_{ij}((u_1,\ldots,u_t)) := u_{ij} :=$ the i-th component of u_j for $i = 1,\ldots,m$ and $j = 1,\ldots,t$. This is the reason we had to double index the variables of Q .) The surjective homomorphism r_t will then induce a closed immersion of affine k-schemes $U^t \hookrightarrow \Omega := \mathrm{Hom}_k(Q,k)$. This in turn induces an inclusion $U[z] \hookrightarrow \Omega$ and it is an easy exercise ([139], page 51) to show that $U[z]$ is dense in Ω .

We can identify $U[z]$ as a subset of U_s the set of input sequences as follows: First note that $U[z]$ is the set of all infinite sequences with finitely many non-zero entries $u = (u_1,\ldots,u_t,0,\ldots)$. Then such a $u \in U[z]$ defines an input sequence $\mathbb{Z} \to U$ also denoted by u, by the rule $u(1-j) := u_j$. Call an arbitrary map $\tilde{f}: U[z] \to Y$ a response map. Then we claim that there is a bijection between constant causal input/output maps f and response maps $\tilde{f}$. Indeed define $\beta: Y_s \to Y$ by $y(t) \longmapsto y(1)$ and let $\sigma^t_{U_s}: U_s \to U_s$ be the composition of the shift σ_{U_s} taken t times. Then to $f: U_s \to Y_s$ we associate $\tilde{f}_f := \beta \circ f | U[z]$ and to any $\tilde{f}: U[z] \to Y$ we associate $f_{\tilde{f}}: U_s \to Y_s$ defined by $f_{\tilde{f}}(u)(t) := \tilde{f}(\sigma^t_{U_s} u)$. It is trivial to check this is a 1-1 correspondence.

We then say that a response map $\tilde{f}: U[z] \to Y$ is polynomial if it is the restriction of a morphism of affine k-schemes $\Omega \to Y$. (Note that $U[z]$ is not an affine k-scheme but only a dense subset of one.) A constant causal input/output map is polynomial if its associated response map is polynomial.

Let Σ be a polynomial system with the assumptions of (6.3). Let $g: U[z] \to X$ be as in (6.4) (iii) and let $\eta: X \to Y$ be the readout map. Then to Σ we get a natural response map $f_\Sigma := \eta \circ g$. Given a response map $\tilde{f}$ we say Σ realizes $\tilde{f}$ if $\tilde{f}_\Sigma = \tilde{f}$. (This can all be formulated in terms of the associated input/output maps of course.)

We can now state the following key result ([139], page 68) which is perhaps the most important result of polynomial realization theory:

Theorem (6.9). For $\tilde{f}: U[z] \to Y$ a polynomial response map, there exists an algebraically canonical realization Σ of $\tilde{f}$ which is unique up to isomorphism in the following sense: Given any other algebraically canonical realization Σ' of $\tilde{f}$ with state space X' , state transition map φ' , readout map η' , there exists an isomorphism $\lambda: X \to X'$ of affine k-schemes such that the diagram

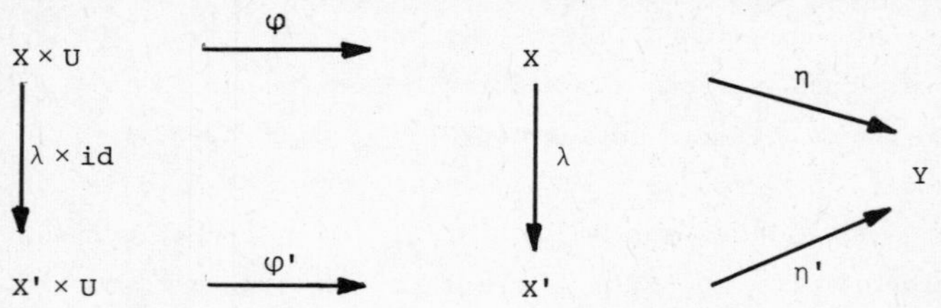

commutes.

Proof. We will indicate the proof of the existence of such an algebraically canonical realization here and leave the reader to look up the proof of the existence of λ in [139], pages 67-68.

We first construct a system $\tilde{\Sigma}$ which is quasi-reachable and which realizes $\tilde{f}$. We let $\tilde{\Sigma}$ have state space Ω (recall from (6.8) this is $\mathrm{Hom}_k(Q,k)$ where Q is the k-algebra of Volterra series), and since by hypothesis $\tilde{f}$ is polynomial it extends to a morphism from $\Omega \to Y$ which we will denote by $\tilde{f}_1$. Then this $\tilde{f}_1$ we will take to be the readout map of the system $\tilde{\Sigma}$. We will take (0) to be the equilibrium point for $\tilde{\Sigma}$. Thus we need only define an appropriate state transition map $\delta: \Omega \times U \to \Omega$. In order to avoid a proliferation of indices we define δ in case $m = 1$ $(U \cong k^m)$. The case $m > 1$ is a simple generalization.

For x a variable, we have that $R_U \cong k[x]$ and to define a morphism $\delta: \Omega \times U \to \Omega$ we need only define a k-algbera homomorphism $\delta^*: Q \to Q \otimes k[x]$. Since $m = 1$, we drop the index i from the z_{ij} and we get that our Volterra series are formal power series in the variables $z_1, z_2, \ldots$. Then if $\alpha \in Q$, we let $\delta^*(\alpha)$ be the Volterra series in the variables $x, z_1, z_2, \ldots$ gotten by substituting x for z_1 and then z_{j-1} for z_j for $j > 1$ in α . This clearly defines a k-algebra homomorphism δ^* whose associated morphism $\delta: \Omega \times U \to \Omega$ we will take to be the state transition map of our system $\tilde{\Sigma}$.

To see that $\tilde{\Sigma}$ realizes $\tilde{f}$ note that by construction the associated reachability map $\tilde{g}$ ((6.4) (iii)) is nothing but the natural inclusion of $U[z] \longrightarrow \Omega$ described in (6.8). Then since Ω is the state space of the system with readout map $\tilde{f}_1: \Omega \to Y$ the extension of the given $\tilde{f}: U[z] \to Y$, we have that $f_{\tilde{\Sigma}} := \tilde{f}_1 \circ \tilde{g} = \tilde{f}$. Also since $U[z]$ is dense in Ω , clearly $\tilde{\Sigma}$ is quasi-reachable.

Finally it is not difficult to show that given any polynomial system Σ with state space X , there exists an algebraically observable system Σ_{obs} with state space X_{obs} and a morphism $\lambda_{obs}: X \to X_{obs}$ such that the image of λ_{obs} is dense in X_{obs} (we say that λ_{obs} is a dominating morphism) and such that the following diagram commutes:

$$\begin{array}{ccccc}
X \times U & \xrightarrow{\ \varphi\ } & X & \searrow^{\eta} & \\
\Big\downarrow{\scriptstyle \lambda_{obs} \times id} & & \Big\downarrow{\scriptstyle \lambda_{obs}} & & Y \\
X_{obs} \times U & \xrightarrow{\ \varphi_{obs}\ } & X_{obs} & \nearrow_{\eta_{obs}} &
\end{array}$$

where φ_{obs} is the state transition map and η_{obs} is the readout map of Σ_{obs} . We say Σ dominates Σ_{obs} . (To explicitly construct X_{obs} and λ_{obs} , we let X_{obs} be the affine k-scheme $Hom_k(A_\Sigma, k)$ where A_Σ is the observation algebra (6.7)(ii). Then λ_{obs} is the morphism $X \to X_{obs}$ which is associated to the inclusion $A_\Sigma \hookrightarrow R_X$. Clearly λ_{obs} is dominating. The required extension of $\varphi: X \times U \to X$ to $\varphi_{obs}: X_{obs} \times U \to X_{obs}$ is carried out in [139], pages 65-66.)

In particular we can construct an algebraically observable system $\tilde{\Sigma}_{obs}$ dominates by $\tilde{\Sigma}$ which must also realize $\tilde{f}$ by the commutativity of the above diagram, and which must be quasi-reachable by the dominating property. This system then satisfies the requirements of the theorem and completes the proof.

Q.E.D.

PART VII. ON THE GEOMETRY OF RATIONAL TRANSFER FUNCTIONS

In this part of the lectures, we would like to introduce the reader to some very pretty classical and modern questions regarding the structure of real and complex strictly proper rational functions in the scalar input/output case. The specific problem is to attempt to parametrize in some natural way the space of strictly proper rational functions of fixed McMillan degree (i.e. the degree of the denominator) n with no common factors. We have seen in Part II, Section 5 how such functions give coordinate-free descriptions of canonical scalar input/output systems.

From a system theoretic point of view this problem is very important in so-called system identification theory. This concerns building a model of a system based on observations of the system in operation. A parametrized class of mathematical models is chosen and then parameters selected so that the model represents the available data about the system in some optimal way. But such a description of system parameters is precisely the question we are asking in the preceding paragraph in the linear scalar input/output case.

The specific problem of describing the topology of the space of rational functions was begun in Brockett [12] and then motivated by Brockett's work, continued by Segal [124] from a homotopy theoretic point of view. However the study of strictly proper real rational functions has a much longer history of course, and has been dealt with classically by such mathematicians as Cauchy [23], Hermite [61] and Hurwitz [72]. They were interested for example in the stability properties of such functions, a question we shall return to in Part VIII of these notes.

We will use some elementary algebraic topology below. Almost all the relevant definitions we will give and try to elucidate. However, lest these notes become too long at certain points we will cheat a bit and refer the reader to the relevant literature. Every definition we make about this topic may be found in the book of Spanier [142]. For less encyclopaedic treatments see Hu [30], Massey [99], and Singer-Thorpe [136].

§1. Cauchy Indices and the Connected Components of $\mathrm{Rat}(n)$

In this section we relate the classical Cauchy index to the connectivity properties of the space of real strictly proper transfer functions as well as give Brockett's neat proof [12] of a classical result due to Hermite-Hurwitz. Our treatment here is based on Brockett [12], Gantmacher, Vol. II [39], and Glover [40].

We begin with a formal definition of the object we want to study:

Real Transfer Functions (1.1).

(i) We set

$$\mathrm{Rat}(n) := \{f/g \mid f,g \in \mathbb{R}[z] \ , \ \text{degree } f < \text{degree } g = n \ , \ g \text{ monic and } f \text{ and } g \text{ have no common factors in } \mathbb{R}[z]\} \ .$$

(ii) From our remarks in Part II, Section 5 we have that Rat(n) may be identified with $\mathcal{M}^c_{1,n,1}$, the quotient space of canonical scalar input/output systems over $\mathbb{R}$.

(iii) We can clearly identify an element

$$\frac{a_{n-1}z^{n-1} + \ldots + a_0}{z^n + b_{n-1}z^{n-1} + \ldots + b_0} \in \mathrm{Rat}(n)$$

with the 2n-tuple $(a_0,\ldots,a_{n-1},b_0,\ldots,b_{n-1})$. Then $\mathrm{Rat}(n) \subset \mathbb{R}^{2n}$ is the complement of the hypersurface defined by the resultant in $\mathbb{R}^{2n}$ and as such is Zariski open.

(iv) We give however $\mathrm{Rat}(n) \subset \mathbb{R}^{2n}$ the induced Euclidean metric topology of $\mathbb{R}^{2n}$. This topology is the same as the compact-open topology with the elements of Rat(n) being regarded as maps from $\mathbb{C} \cup \{\infty\} \cong S^2 \to S^2$ (S^2: = real 2-sphere). (Recall that the compact-open topology on the space of all continuous functions from $X \to Y$ (X,Y topological spaces) is the topology generated by the subbase $\{\langle K;U\rangle\}$ where

$$\langle K;U\rangle := \{f\colon X \to Y \mid f \text{ is continuous, } f(K) \subset U \ , \ K \text{ is compact, and } U \text{ is open}\} \ .)$$

(v) If the resultant locus is defined by the equation $\psi = 0$, then via the standard trick of introducing an extra variable y , we have that Rat(n) may be regarded to be the real hypersurface of degree 2n in $\mathbb{R}^{2n+1}$ defined by $\psi y - 1 = 0$. Using Morse theory, one may then make some computations about homology of Rat(n) . See Milnor [101].

We want to compute the number of connected components of Rat(n) . It is a nice result of Brockett [12] that these are precisely classified by the Cauchy index [23] which we now define:

Definition (1.2). Let $h \in \mathrm{Rat}(n)$ and let p be a real pole. Then the index of h at p , denoted by $\mathrm{ind}_p h$, is defined to be +1 if h changes from $+\infty$ to $-\infty$, -1 if h changes from $-\infty$ to $+\infty$, and 0 if h doesn't change sign while passing through p . The Cauchy index is $\deg h := \sum_{p \text{ real pole}} \mathrm{ind}_p(h)$.

Remark (1.3). There are several equivalent characterizations of the Cauchy index which we would like to give here. If h has no repeated poles, and if one expands h in terms of partial fractions, then it is immediate that the Cauchy index is the number of real poles with positive residues minus the number of real poles with negative residues.

A more topological formulation may be given as follows: An element $h \in \mathrm{Rat}(n)$ defines a continuous mapping from $\mathbb{C} \cup \{\infty\} \to \mathbb{C} \cup \{\infty\}$ (regarded as the Riemann sphere S^2). Moreover since h is real, it sends $\mathbb{R} \cup \{\infty\} \to \mathbb{R} \cup \{\infty\}$ and $\mathbb{R} \cup \{\infty\}$ is homeomorphic to S^1. Now as a map from $S^2 \to S^2$, h has a degree (for the definition see Hu [70], page 12) and this degree is clearly n. Moreover, from the realization theory of Part VI it should be obvious that if we expand $h = \sum_{i \geq 1} c_i z^{-i}$ as a formal Laurent series, and consider the associated $n \times n$ Hankel $\mathcal{H}(h)$, then rank $\mathcal{H}(h) = n$.

Now $h | \mathbb{R} \cup \{\infty\}$ also has a degree, the so-called "winding number". It is again obvious (see e.g. Ahlfors [1]) that the winding number is exactly deg h, the Cauchy index of h. Note moreover that $\mathcal{H}(h)$ being a real symmetrix matrix implies that all the eigenvalues of $\mathcal{H}(h)$ are real and we can define the _signature_ of $\mathcal{H}(h)$ to be the difference between the number of positive and negative eigenvalues. Then the only non-trivial result of this discussion is the fact due to Hermite [61] and Hurwitz [72], which states that the signature of $\mathcal{H}(h)$ is precisely the Cauchy index. We will indicate the modern proof given in [12] of this important result in (1.5) below.

We first have however the following result also from [12]:

Proposition (1.4). Rat(n) _has precisely_ n+1 _connected components given by_ $\text{Rat}(p,q) := \{h \in \text{Rat}(n) \mid \deg h = p-q\}$ _for each pair of positive integers_ (p,q) _such that_ $p+q = n$. _In particular_ deg h _is congruent to_ n _modulo_ 2 _for each_ $h \in \text{Rat}(n)$.

Proof. We will sketch here Brockett's original proof. In Section 3 we will re-prove this using a much stronger result of Segal [124].

First it is obvious that $\deg: \text{Rat}(n) \to \mathbb{Z}$ is continuous. Moreover deg at least takes on the values $-n, -n+2, \ldots, n-2, n$ since for each j,
$$\deg\left(\sum_{s=j}^{n-j} \frac{j}{z+j} \right) = n-2j .$$ Thus Rat(n) has at least n+1 connected components and hence we will be done once we show that Rat(p,q) is connected for all $p + q = n$ $p,q \geq 0$. One can show however through a series of deformations of the roots and poles of a given $h \in \text{Rat}(p,q)$ (done in such a way as to avoid common factors), that h can be put in a certain normalized form ([12], page 451, Lemma 6), and this implies precisely that Rat(p,q) is arcwise connected.

Q.E.D.

Corollary (1.5). (Hermite-Hurwitz). Let $h \in \text{Rat}(p,q)$, _and let_ $\mathcal{H}(h)$ _be the associated_ $n \times n$ _Hankel matrix. Then the signature of_ $\mathcal{H}(h) = p-q$.

Proof. First we claim that the values of signature $\mathcal{H}(h)$ and deg h are 1-1 correspondence. Indeed to prove this, we must show that for $h_1, h_2 \in \text{Rat}(n)$ with signature $\mathcal{H}(h_1) \neq$ signature $\mathcal{H}(h_2)$, h_1 and h_2 lie in different connected components of Rat(n). But if h_1 and h_2 lay in the same connected component, then we could find a connected path from h_1 and h_2 and hence a deformation from $\mathcal{H}(h_1)$ to $\mathcal{H}(h_2)$. But since $\mathcal{H}(h_1)$ and $\mathcal{H}(h_2)$ have different numbers of positive eigenvalues (since their signatures are different), some of these eigenvalues would go to zero which would mean that the corresponding rational function would have common factors in its numerator and denominator, contradicting the definition of Rat(n) .

We are reduced to showing now that $h_k := \sum_{j=-k}^{n-k} \frac{j}{z+j}$ (so that $h_k \in \mathrm{Rat}(n-k,k)$) is such that signature of $\mathcal{H}(h_k) = n-2k$ for each $k = 0,1,\ldots,n$. But we can canonically realize the strictly proper rational functions h_k as follows:

$$F_k := \begin{pmatrix} k & & & & & 0 \\ & k-1 & & & & 0 \\ & & \ddots & & & \\ & & & 1 & & \\ 0 & 0 & & & -1 & \\ & & & & & -n+k \end{pmatrix} \qquad (F_k \text{ is diagonal})$$

$$G_k := \begin{pmatrix} \sqrt{k} \\ \sqrt{k-1} \\ \vdots \\ 1 \\ 1 \\ \vdots \\ \sqrt{n-k} \end{pmatrix}$$

$$H_k := (-\sqrt{k}\ \ -\sqrt{k-1}\ \ldots\ -1\ \ 1\ \ldots\ \sqrt{n-k}) .$$

It is trivial to verify that for each $k = 0,1,\ldots,n$, (F_k,G_k,H_k) is indeed a canonical realization of h_k and that the associated Hankel has signature $n-2k$.

Q.E.D.

Remark (1.6). In general one can set for any $m,p \geq 1$

$$\mathrm{Rat}^{p,m}(n) := \{p \times m \text{ strictly proper real rational functions of McMillan degree } n\}$$

where, as we recall the McMillan degree of such a function is the dimension of a canonical realization.

Glover [40] has shown that for $m > 1$ or $p > 1$ (i.e. the multi-input or multi-output case), $\mathrm{Rat}^{p,m}(n)$ is connected (taken with the obvious topology). Indeed in this case the space of real non-canonical systems of dimension n with m inputs and p outputs has codimension ≥ 2 in the space of all such systems, since these are defined by the vanishing of at least two determinantel equations. Elementary topological considerations and realization theory imply that $\mathrm{Rat}^{p,m}(n)$ is connected.

Explicitly using the notation of Part IV, for a non-canonical system (F,G,H) of dimension n we have conditions $\mathrm{rank}\, R(F,G) < n$ or $\mathrm{rank}\, R(F^t,H^t) < n$. Suppose $m > 1$. Then in the non-completely reachable case we have at least two nice selections I , J and at least two equations $\det R(F,G)_I = \det R(F,G)_J = 0$ defining the condition of non-reachability. It is not difficult (for details see [40]) to show that in general these equations are independent from which the claim about the codimension follows immediately.

§2. Complex Transfer Functions

In this section we will set the background for deducing the main results about the topology of Rat(n) which we will discuss in Section 3. We shall also prove a very nice result of J. Jones [124] which will be very important for us in Part VIII of these notes.

Following Segal [124] in order to deduce the topological properties of Rat(n) it will be essential to first study the complex case. We will begin therefore with a definition of a complexified Rat(n) as well as some elementary definitions from homotopy theory which we will need to understand Segal's work.

Definitions-Remarks (2.1).

(i) We set

(a) $\mathrm{Rat}_{\mathbb{C}}(n) := \{f/g \mid f,g \in \mathbb{C}[z]$, degree f < degree g = n , g monic, f and g have no common factors in $\mathbb{C}[z]\}$.

(b) $\mathrm{Rat}'(n) := \{f/g \mid f,g \in \mathbb{R}[z]$, degree f = degree g = n , f and g monic, f and g have no common factors in $\mathbb{R}[z]\}$.

(c) $\mathrm{Rat}'_{\mathbb{C}}(n) := \{$same as (b) with $f,g \in \mathbb{C}[z]\}$.

(ii) As in (1.1), $\mathrm{Rat}_{\mathbb{C}}(n)$ is considered to be a Euclidean open subset of $\mathbb{C}^{2n}$, defined by the complement of the resultant locus. Of course $\mathrm{Rat}_{\mathbb{C}}(n)$ is connected. (The Euclidean topology on $\mathrm{Rat}_{\mathbb{C}}(n)$ again coincides with the compact-open topology.)

(iii) In studying the topological properties of Rat(n) and $\mathrm{Rat}_{\mathbb{C}}(n)$ it will be sometimes more convenient to study Rat'(n) and $\mathrm{Rat}'_{\mathbb{C}}(n)$. Note that $\mathrm{Rat}(n) \cong \mathrm{Rat}'(n)$ via the mapping $h \longmapsto h+1$, and similarly $\mathrm{Rat}_{\mathbb{C}}(n) \cong \mathrm{Rat}'_{\mathbb{C}}(n)$. The notation Rat'(p,q) (p+q = n) should be self-explanatory.

We now give a very brief sketch of the relevant concepts from algebraic topology needed to understand the results below. This sketch is just intended to give the reader uninitiated to the mysteries of algebraic topology a flavor of the subject and to motivate what follows. However for the full picture the reader will have to consult the references on topology mentioned above.

Some Algebraic Topology (2.2).

(i) Let $\lambda_1, \lambda_2 \colon X \to Y$ be continuous maps of topological spaces X and Y . Then λ_1 and λ_2 are homotopic (denoted by $\lambda_1 \simeq \lambda_2$) if there exists a continuous map $\lambda \colon X \times [0,1] \to Y$ such that for every $x \in X$, $\lambda(x,0) = \lambda_1(x)$ and $\lambda(x,1) = \lambda_2(x)$.

Equivalently, λ_1 and λ_2 are homotopic if they can be connected by an arc in the space of all continuous functions from X to Y (this space being given the compact-open topology).

(ii) It is immediate that $\simeq$ is an equivalence relation.

(iii) Let $\lambda_1, \lambda_2 \colon X \to Y$ be as in (i), and let $S \subset X$ be an arbitrary subset. Then λ_1 and λ_2 are homotopic relative to S if there exists a continuous $\lambda \colon X \times [0,1] \to Y$ satisfying the same conditions as in (i) with the additional condition that $\lambda(s,t) = \lambda_1(s) = \lambda_2(s)$ for all $s \in S$, $t \in [0,1]$.

(iv) Let B_n be the boundary of $[0,1]^n$ (i.e. the Cartesian product of the closed unit interval $[0,1]$ taken n times; this is called the n-cube in $\mathbb{R}^n$). Then define for X a topological space and $x \in X$ a point,

$$\pi_n(X,x) := \{\text{relative homotopy classes of maps } \varphi\colon [0,1]^n \to X \text{ such that } \varphi(B_n) = x\}.$$

Note that the term "relative homotopy class" refers to the fact that we take all homotopies relative to B_n. Given $\varphi_1, \varphi_2 \colon [0,1]^n \to X$ with $\varphi_1(B_n) = \varphi_2(B_n) = x$ we can define

$$(\varphi_1 + \varphi_2)(x_1,\ldots,x_n) := \begin{cases} \varphi_1(2x_1, x_2, \ldots, x_n) & \text{for } 0 \leq x \leq \frac{1}{2} \\ \varphi_2(2x_1 - 1, x_2, \ldots, x_n) & \text{for } \frac{1}{2} \leq x_1 \leq 1 \end{cases}.$$

Then from [70], this defines a group structure on $\pi_n(X,x)$ and $\pi_n(X,x)$ is called the n-th homotopy group of X.

If we give the space of maps from $[0,1]^n \to X$ which map B_n to x the compact-open topology, then $\pi_n(X,x)$ is just the set of all path components of this space. Note for X arcwise connected $\pi_n(X,x)$ is independent of x. For $n = 1$, $\pi_1(X,x)$ is called the fundamental group of X.

(v) Given $\lambda \colon X \to Y$ any continuous map, it is easy to show ([70]) that λ induces a group homomorphism $\lambda_* \colon \pi_n(X,x) \to \pi_n(Y,\lambda(x))$ for each $x \in X$. Now let $\lambda_1, \lambda_2 \colon X \to Y$ be homotopic relative to $\{x\} \subset X$. Then immediately $\lambda_{1*} = \lambda_{2*}$. Therefore the groups $\pi_n(X,x)$ are invariants of the based homotopy class of the space X.

(vi) The groups $\pi_n(X,x)$ "measure" the connectivity properties of the space X. For example, for X arcwise connected, X is simply connected (i.e. has no "holes") if $\pi_1(X,x) \cong \{1\}$. A nice result (see Massey [99], pages 68-74) is that $\pi_1(S^1,x) \cong \mathbb{Z}$ where S^1 denotes the circle. For $S^1 \vee S^1$ (two circles joined at a point) $\pi_1(S^1 \vee S^1, x) \cong \mathbb{Z} * \mathbb{Z}$ (the free product of Z and $\mathbb{Z}$). Intuitively then, for each independent loop of an arcwise connected space X, we get one copy of $\mathbb{Z}$ in $\pi_1(X,x)$. Since we are interested in studying the connectivity properties of Rat(n), this motivates our looking at these homotopy groups.

(vii) The proof that $\pi_1(S^1,x) \cong \mathbb{Z}$ makes use of the fact that $\exp \colon \mathbb{R} \to S^1$ with $\exp(r) := e^{2\pi i r}$ defines $\mathbb{R}$ as a covering space of S^1. In general given a continuous map $p \colon \tilde{X} \to X$ of topological spaces, $(\tilde{X}, p)$ is a covering space of X if p is surjective and for each $x \in X$, there exists an open neighborhood U containing x such that $p^{-1}(U)$ is a disjoint union of open sets each mapped homeomorphically to U via p . $(\widetilde{X},p)$ is universal if $\widetilde{X}$ is simply connected. (exp: $\mathbb{R} \to S^1$ is an example of a universal covering space.)

(viii) Given $p: \widetilde{X} \to X$ as in (vii), a covering transformation is a homeomorphism $h: \widetilde{X} \to \widetilde{X}$ such that $p \circ h = p$. Clearly the set of all these covering transformations is a group. Notice that if U is as in (vii), then h permutes the copies of U in $p^{-1}(U)$. Moreover if $p_*\pi_1(\widetilde{X},\widetilde{x})$ is a normal subgroup of $\pi_1(X,p(\widetilde{x}))$, i.e. $(\widetilde{X},p)$ is regular, then one can show ([99]) that the group of covering transformations is isomorphic to $\pi_1(X,\pi(\widetilde{x})/p_*\pi_1(X,x)$.

(ix) Let $\widetilde{\mathbb{C}}^{(n)} := \{(z_1,\ldots,z_n) \in \mathbb{C}^n \mid z_i \neq z_j \text{ for } i \neq j\}$. Then the symmetric group Σ_n on n-letters acts in the natural way on $\widetilde{\mathbb{C}}^{(n)}$ by permuting the z_i , and we can consider the quotient space $\mathbb{C}^{(n)} := \widetilde{\mathbb{C}}^{(n)}/\Sigma_n$. Thus $\mathbb{C}^{(n)}$ is the space of unordered n-tuples of distinct points of $\mathbb{C}$. The n-th Braid group is $Br_n := \pi_1(\mathbb{C}^{(n)})$. (We drop the explicit reference to the base point $x \in \mathbb{C}^{(n)}$ since $\mathbb{C}^{(n)}$ is clearly arcwise connected; see (iv).)

(x) There is a whole theory associated to Br_n . For complete details see the monograph of Birman [9]. One easy fact that we will need below is that the natural projection $p: \widetilde{\mathbb{C}}^{(n)} \to \mathbb{C}^{(n)}$ is a regular covering space projection, with group of covering transformations isomorphic to Σ_n . Thus from (ix), $\pi_1(\mathbb{C}^{(n)})/p_*\pi_1(\widetilde{\mathbb{C}}^{(n)}) \cong \Sigma_n$.

(xi) Finally given a triple (X,A,x) consisting of a topological space X , a non-empty subset $A \subset X$, and a point $x \in A$ one can define relative homotopy groups $\pi_n(X,A,x)$. Briefly if $[0,1]^{n-1} \subset [0,1]^n$ is a face of the n-cube, and J denotes the union of the other faces, then one can consider maps $f: [0,1]^n \to X$ which take $[0,1]^{n-1}$ to A , and J to x . The totality of homotopy classes of these maps (defined relative to the obvious system; see [70], pages 15 and 110) is denoted by $\pi_n(X,A,x)$.

We are ready to state the fundamental result from G. Segal's powerful paper [124]. The proof is beyond the scope of these notes and we state it without proof (the statement in [124] is (1.1), page 39):

Theorem (2.3). Let $K_n = \{$maps $S^2 \to S^2$ which have degree n and which take ∞ to $0\}$. (We give K_n the compact-open topology.) Then the inclusion $Rat_{\mathbb{C}}(n) \hookrightarrow K_n$ is a homotopy equivalence up to dimension n .

Remark (2.4). It is (2.3) which will allow us to compute the homotopy groups of Rat(n) in Section 3.

We conclude this section with a result of J. Jones also proven in [124], page 63. Since this result will be quite important in Part VIII, Section 4, and since the techniques are themselves interesting, we would like to give the proof in some detail:

Theorem (2.5). $\pi_1(\mathrm{Rat}'_{\mathbb{C}}(n)) \cong \mathbb{Z}$.

Proof. Let $U := \{f/g \in \mathrm{Rat}'_{\mathbb{C}}(n) \mid g$ has n distinct roots$\}$. Clearly the real codimension of $\mathrm{Rat}'_{\mathbb{C}}(n)-U$ in $\mathrm{Rat}'_{\mathbb{C}}(n)$ is 2 . Now from Hu [70], page 115, associated to the couple $(\mathrm{Rat}'_{\mathbb{C}}(n),U)$ we obtain an exact sequence of groups

$$\cdots \to \pi_1(u) \to \pi_1(\mathrm{Rat}'_{\mathbb{C}}(n)) \to \pi_1(\mathrm{Rat}'_{\mathbb{C}}(n),U) \to \cdots$$

and by [70], page 112, $\pi_1(\mathrm{Rat}'_{\mathbb{C}}(n),U) = 0$. Thus $\pi_1(U) \to \pi_1(\mathrm{Rat}'_{\mathbb{C}}(n))$ is surjective.

Using the notation of (2.2), (ix) and (x), we have that there exists a natural mapping $U \to \mathbb{C}^{(n)}$ gotten by sending $f/g \in U$ to the unordered n-tuple of distinct points defined by the roots of g . This mapping is clearly a fibration with fiber $(\mathbb{C}^*)^n$. Moreover it is clear that this fibration has a section (e.g. if we write elements of $\mathbb{C}^{(n)}$ as formal sums $\beta_1 + \cdots + \beta_n$ then we can define a section via

$$\beta_1 + \cdots + \beta_n \longmapsto \frac{\left(\prod_{i=1}^{n} (z-\beta_i)\right)-1}{\prod_{i=1}^{n} (z-\beta_i)} \quad .)$$

Given such a fibration $\mathbb{C}^{*n} \to U \to \mathbb{C}^{(n)}$ one can construct an exact sequence of homotopy groups ([70], page 152)

$$\cdots \to \pi_2(\mathbb{C}^{(n)}) \to \pi_1(\mathbb{C}^{*n}) \to \pi_1(U) \to Br_n \to 1$$

and from the fact that $p: \tilde{\mathbb{C}}^{(n)} \to \mathbb{C}^{(n)}$ is a covering space we have ([70], page 154) that $\pi_2(\mathbb{C}^{(n)}) \cong \pi_2(\tilde{\mathbb{C}}^{(n)}) \cong \{1\}$. Hence we have that

$$1 \to \pi_1(\mathbb{C}^{*n}) \to \pi_1(U) \to Br_n \to 1$$

is exact. The existence of a section for $U \to \mathbb{C}^{(n)}$ implies that $\pi_1(U)$ is the semidirect product of Br_n and $\pi_1(\mathbb{C}^{*n}) \cong \mathbb{Z}^n$ (see Hu [70], page 152 and Humphreys [71], page 61). In point of fact we can write the action of Br_n on $\mathbb{Z}^n$ quite explicitly. Indeed from (2.2)(x) we have that $Br_n/p_*\pi_1(\tilde{\mathbb{C}}^{(n)}) \cong \Sigma_n$. Thus we have a homomorphism of groups $Br_n \to \Sigma_n$, and Br_n acts on $\mathbb{Z}^n$ by permutations via this homomorphism.

Next we set

$V := \{f/g \in U \mid$ roots of f are in the open upper half plane H^+ and the roots of g are in the open lower half plane $H^-\}$.

Let $H_n^- :=$ {unordered n-tuples of distinct points of H^-} (see (2.2)(ix)). Then one has the natural mapping $V \to H_n^-$ defined by sending $f/g \in V$ to the roots of g . Note moreover that the fiber of $V \to H_n^-$ is $(H^+)^n$ which is contractible. It is clear that H_n^- is homotopy equivalent to $\mathbb{C}^{(n)}$ since H^- and $\mathbb{C}$ are homotopy equivalent. Therefore from the exact homotopy sequence associated to the fibration $(H^+)^n \to V \to H_n^-$ ([70], page 152), we see that $\pi_1(V) \xrightarrow{\sim} \pi_1(H_n^-) \cong Br_n$.

Now we claim that $V \hookrightarrow \mathrm{Rat}'_{\mathbb{C}}(n)$ is homotopic to a constant. Indeed to see this choose fixed distinct $a_1,\dots,a_n \in H^+$, and fixed distinct $b_1,\dots,b_n \in H^-$. Let

$$h = \frac{\prod_{i=1}^{n} (z-\alpha_i)}{\prod_{i=1}^{n} (z-\beta_i)}$$

be an arbitrary element of V. Then we can deform h to $\frac{\Pi(z-a_i)}{\Pi(z-b_i)}$ in $\mathrm{Rat}'_{\mathbb{C}}(n)$ via the homotopy

$$h_t = \frac{\Pi(z-ta_i-(1-t)\alpha_i)}{\Pi(z-tb_i-(1-t)\beta_i)} .$$

Next we have seen that $\pi_1(U)$ is the semidirect product $Br_n \times \mathbb{Z}^n$ and that $Br_n \times \mathbb{Z}^n \to \pi_1(\mathrm{Rat}'_{\mathbb{C}}(n))$ surjectively. But since $Br_n \cong \pi_1(V)$ and V is homotopic to a constant in $\mathrm{Rat}'_{\mathbb{C}}(n)$, we have that Br_n is contained in the kernel of $Br_n \times \mathbb{Z}^n \to \pi_1(\mathrm{Rat}'_{\mathbb{C}}(n))$. This means that $\pi_1(\mathrm{Rat}'_{\mathbb{C}}(n))$ is isomorphic to a quotient of $Br_n \times \mathbb{Z}^n$ by a normal subgroup containing Br_n and by direct computation (we have the action of Br_n on $\mathbb{Z}^n$ explicitly) one sees the largest such quotient is $\mathbb{Z}$. We must therefore show now that $\pi_1(\mathrm{Rat}'_{\mathbb{C}}(n))$ cannot be smaller than $\mathbb{Z}$.

To do this we consider the resultant map $R\colon \mathrm{Rat}'_{\mathbb{C}}(n) \to \mathbb{C}^*$. Explicitly if $h = (\Pi(z-\alpha_i))/(\Pi(z-\beta_i)) \in \mathrm{Rat}'_{\mathbb{C}}(n)$, $R(h) := \prod_{i,j} (\alpha_i-\beta_j)$ (see Lang [91]). Now it is rather clear that $R\colon \mathrm{Rat}'_{\mathbb{C}}(n) \to \mathbb{C}^*$ is a fiber bundle. To see this note that if

$$h = \frac{z^n+a_1z^{n-1}+\dots+a_n}{z^n+b_1z^{n-1}+\dots+b_n}$$

then $R(h)$ is a polynomial in $a_1,\dots,a_n$, $b_1,\dots,b_n$. If we consider a_j and b_j to be of weight j $(j = 1,\dots,n)$, then $R(h)$ is a homogeneous polynomial of weight n^2. Now $\mathbb{C}^*$ acts on $\mathrm{Rat}'_{\mathbb{C}}(n)$ by

$$c\cdot(a_1,\dots,a_n,\ b_1,\dots,b_n) = (ca_1,\dots,c^na_n,cb_1,\dots,c^nb_n)$$

and thus $R(c\cdot h) = c^{n^2}R(h)$. This all implies that $R\colon \mathrm{Rat}'_{\mathbb{C}}(n) \to \mathbb{C}^*$ is a fiber bundle with structure group, the group of n^2 roots of unity. Consequently again from [70], page 152, we see that $\pi_1(\mathrm{Rat}'_{\mathbb{C}}(n)) \to \pi_1(\mathbb{C}^*)$ surjectively and since $\pi_1(\mathbb{C}^*) \cong \mathbb{Z}$, this completes the proof.

Q.E.D.

Remarks (2.6).

(i) The above proof implies that a generator of $\pi_1(\mathrm{Rat}'_{\mathbb{C}}(n))$ is defined by a loop which moves a pole once around a zero of a given $h \in \mathrm{Rat}'_{\mathbb{C}}(n)$. See also Part VIII, (4.10).

(ii) In the fiber bundle $R\colon \mathrm{Rat}'_{\mathbb{C}}(n) \to \mathbb{C}^*$, since $\pi_1(\mathrm{Rat}'_{\mathbb{C}}(n)) \cong \pi_1(\mathbb{C}^*)$ and $\pi_2(\mathbb{C}^*) = 0$, from the exact sequence

$$\dots \to \pi_2(\mathbb{C}^*) \to \pi_1(R^{-1}(1)) \to \pi_1(\mathrm{Rat}'_{\mathbb{C}}(n)) \to \pi_1(\mathbb{C}^*) \to 1$$

we see that $\pi_1(R^{-1}(1)) \cong \{1\}$, i.e. the fibers of $R\colon \mathrm{Rat}'_{\mathbb{C}}(n) \to \mathbb{C}^*$ are simply connected.

(iii) This remark is addressed to readers who know something about integral cohomology or integral homology ([142]). Since we shall not use it in the sequel it can be safely skipped.

We have seen in (ii) that a fiber F of $R: \mathrm{Rat}'_{\mathbb{C}}(n) \to \mathbb{C}^*$ is simply connected. In such circumstances one can compute the integral homology (or integral cohomology) of $\mathrm{Rat}'_{\mathbb{C}}(n)$ if one knows the integral homology of the fiber. Indeed up to homotopy we can regard the inclusion $F \longrightarrow \mathrm{Rat}'_{\mathbb{C}}(n)$ as an infinite cyclic covering $(\pi_1(\mathrm{Rat}'_{\mathbb{C}}(n))/\pi_1(F) \cong \mathbb{Z})$ and $\pi_1(\mathrm{Rat}'_{\mathbb{C}}(n))$ acts on $H_k(F) := H_k(F, \mathbb{Z})$. Then using the Serre spectral sequence (see Mislin [102]) one has that

$$(*) \quad 0 \to (H_k(F))_{\mathbb{Z}} \to H_k(\mathrm{Rat}'_{\mathbb{C}}(n)) \to (H_{k-1}(F))^{\mathbb{Z}} \to 0$$

where $(H_k(F))_{\mathbb{Z}} :=$ largest quotient module of $H_k(F)$ on which $\mathbb{Z}$ acts trivially, and $(H_{k-1}(F))^{\mathbb{Z}} := \mathbb{Z}$-invariant part. Now Segal [124], page 61 shows that $\mathrm{Rat}'_{\mathbb{C}}(n)$ is a nilpotent space up to dimension n. Hence for $k < n$ we have that

$$(**) \quad 0 \to H_k(F) \to H_k(\mathrm{Rat}'_{\mathbb{C}}(n)) \to H_{k-1}(F) \to 0 .$$

§3. The Topology of Rat(n)

We are now ready to prove the fundamental results concerning the topology of Rat(n). Our treatment here is based on Segal [124]. We begin with the following formalism:

Divisors (3.1).

(i) Given a Riemann surface S (see Part II, Section 4), we let $\mathrm{Div}(S) :=$ free abelian group generated by the points of S. Thus a typical element of $\mathrm{Div}(S)$ is a formal finite sum $\Sigma n_P P$ where $n_P \in \mathbb{Z}$, $P \in S$. The elements of $\mathrm{Div}(S)$ are called divisors.

(ii) We have a natural group homomorphism $\deg: \mathrm{Div}(S) \to \mathbb{Z}$ defined by setting $\deg(\Sigma n_P P) := \Sigma n_P$. This homomorphism is of course called the degree.

(iii) A divisor $D \in \mathrm{Div}(S)$ is positive or effective if $D = \Sigma n_P P$ where all the $n_P \geq 0$.

(iv) For p,q non-negative integers, we set $\mathrm{Div}_{p,q}(S) := \{(D_1,D_2) \mid D_1,D_2$ positive divisors, such that D_1 and D_2 are disjoint and $\deg D_1 = p$, $\deg D_2 = q\}$.

(v) Consider the space

$\mathrm{Rat}_{p,q}(\mathbb{C}) := \{f/g \mid f,g \in \mathbb{C}[z]$, degree $f = p$, $\deg g = q$, f and g monic f and g have no common factors in $\mathbb{C}[z]\}$.

Clearly $h \in Rat_{p,q}(\mathbb{C})$ is uniquely determined by its roots and poles and hence in the obvious way we have an isomorphism $Rat_{p,q}(\mathbb{C}) \xrightarrow{\approx} Div_{p,q}(\mathbb{C})$. We give $Div_{p,q}(\mathbb{C})$ the topology induced from that of $Rat_{p,q}(\mathbb{C})$.

(vi) Note that $Rat'_{\mathbb{C}}(n) = Rat_{n,n}(\mathbb{C}) \cong Div_{n,n}(\mathbb{C})$. Moreover if $p \geq q$, then $Div_{p,q}(\mathbb{C}) \cong Div_{q,q}(\mathbb{C}) \times \mathbb{C}^{p-q}$ since if $f/g \in Rat_{p,q}(\mathbb{C})$, we can uniquely write $f/g = u+r/g$ where u is a monic polynomial of degree $p-q$ and $r/g \in Rat'_{\mathbb{C}}(q)$.

With this formalism we can prove the following very nice result of Segal [124], page 638 which generalizes (1.4) above:

Proposition (3.2). $Rat'(p,q) \xrightarrow{\approx} Div_{p,q}(\mathbb{C})$. In particular $Rat(p,q) \simeq Rat'(p,q)$ is connected.

Proof. Note that any $h \in Rat'(p,q)$ is determined by the divisor D associated to the set $h^{-1}(i)$. Indeed D determines $\bar{D}$, and since h is real we have therefore that D determines the function $\tilde{h}(z) = (h(z)-i)/(h(z)+i)$ (since D is the divisor of zeroes and $\bar{D}$ the divisor of poles of $\tilde{h}(z)$), and clearly $\tilde{h}$ determines h. Let $H^+ :=$ open upper half plane, $H^- :=$ open lower half plane. Then since h is real, D contains no real values and hence $D = D_1+D_2$ where $D_1 \in Div(H^+)$, $D_2 \in Div(H^-)$. Thus the divisor of roots of $\tilde{h}$ in H^+ is D_1, and the divisor of poles of $\tilde{h}$ in H^+ is $\bar{D}_2$. But if we restrict $\tilde{h}$ to the boundary of H^+, we see that the resulting winding number must be exactly $p-q$ since the Cauchy index of h is $p-q$ and the map $(z-i)/(z+i)$ is an automorphism of the Riemann sphere. But by Cauchy's theorem [1] this winding number is also equal to $\deg D_1 - \deg \bar{D}_2$ i.e. $(D_1,\bar{D}_2) \in Div_{p,q}(H^+)$. Thus we have shown $Rat'(p,q) \xrightarrow{\approx} Div_{p,q}(H^+)$. But it is easy to see $Div_{p,q}(H^+) \cong Div_{p,q}(\mathbb{C})$.

Q.E.D.

Corollary (3.3). Let $\alpha := \min(p,q)$. Then
$$Rat(p,q) \cong Rat(\alpha,\alpha) \times \mathbb{R}^{2|p-q|}.$$

Proof. Since $Rat(p,q) \simeq Rat(q,p)$ we can clearly assume $p \geq q$. Then we have by (3.2)

$$\begin{aligned} Rat(p,q) &\simeq Div_{p,q}(\mathbb{C}) \\ &\cong Div_{q,q}(\mathbb{C}) \times \mathbb{C}^{p-q} \quad \text{(by (3.1) (vi))} \\ &\simeq Rat(q,q) \times \mathbb{R}^{2(p-q)} . \end{aligned}$$

Q.E.D.

Definitions (3.4).

(i) We set

$K_{n,r} :=$ {equivariant maps $S^2 \to S^2$ of degree n taking ∞ to 0 and of degree r on the equator S^1}.

$K_{n,r}$ is of course given the compact-open topology.

(ii) Note that there is a natural inclusion

$$Rat(p,q) \hookrightarrow K_{p+q,p-q} .$$

(iii) In general for $X' \subset X$, $Y' \subset Y$, the symbol $f: (X,X') \to (Y,Y')$ denotes a map $f: X \to Y$ such that $f(X') \subset Y'$.

(iv) Let X be a topological space, $x \in X$. Then the <u>loop space</u> of X , denoted by ΩX , is defined to be the space of continuous functions $f: ([0,1],\{0,1\}) \to (X,x)$ given the compact-open topology. This notion is fundamental in homotopy theory. For example, ΩX is a homotopic-theoretic group i.e. a so-called <u>H-group</u>. For details see Spanier [142], pages 37-39.

(v) The double loop space $\Omega^2 S^2$ has infinitely many connected components, each defined by the degree of a map from $S^2 \to S^2$. All these components are homotopy equivalent (this follows from the H-group structure on $\Omega^2 S^2$) and hence by slight abuse of notation the symbol $\Omega^2 S^2$ will also denote any connected component of $\Omega^2 S^2$.

We now have the following:

<u>Lemma (3.5).</u> <u>$K_{n,r}$ and $\Omega^2 S^2$ are homotopy equivalent.</u>

<u>Proof.</u> Regarding S^2 as the Riemann sphere $(\mathbb{C} \cup \{\infty\})$, $K_{n,r}$ consists of the set of maps $(S^2,S^1) \to (S^2,S^1)$ which have degree n on S^2 , degree r on S^1 $(\mathbb{R} \cup \{\infty\})$, take ∞ to 0 , and which commute with conjugation. Thus $K_{n,r}$ is equivalent to the space of maps $(e_2,S^1) \to (S^2,S^1)$ of degree r on S^1 where e_2 denotes the closed upper hemisphere of S^2 . But this space of maps is homotopy equivalent to maps of fixed degree $(e_2,S^1) \to (S^2,\cdot)$ where $\cdot$ denotes a point, since the space of based maps of fixed degree from $S^1 \to S^1$ is contractible and S^1 is contractible on S^2 . But this latter space of maps is homotopy equivalent to the space of maps $(S^2,\cdot) \to (S^2,\cdot)$ of fixed degree. Indeed one has the natural map $\varphi: (e_2,S^1) \to (S^2,\cdot)$ which just pinches S^1 to a point. Then given $g: (S^2,\cdot) \to (S^2,\cdot)$ we get $g \circ \varphi: (e_2,S^1) \to (S^2,\cdot)$. Thus we see $K_{n,r}$ is homotopy equivalent to $\Omega^2 S^2$ as required.

Q.E.D.

<u>Remark (3.6).</u> Segal [124], page 64, (7.2), proves that the inclusion $\mathrm{Rat}(p,q) \hookrightarrow K_{p+q,p-q}$ is a homotopy equivalence up to dimension $\min(p,q)$. For our purposes however the following weaker result will suffice:

<u>Theorem (3.7).</u> $\mathrm{Rat}(p,q)$ <u>and</u> $K_{p+q,p-q}$ <u>have the same homotopy type up to dimension</u> $\min(p,q)$.

<u>Proof.</u> Since $\mathrm{Rat}(p,q)$ is homeomorphic to $\mathrm{Rat}(q,p)$ and $K_{p+q,p-q}$ is homeomorphic to $K_{q+p,q-p}$, we can clearly assume that $p \geq q$. But from (3.1)(vi) and (3.2) we have that $\mathrm{Rat}(p,q) \cong \mathrm{Div}_{p,q}(\mathbb{C}) \cong \mathrm{Div}_{q,q}(\mathbb{C}) \times \mathbb{C}^{p-q}$. Now from (2.3) we have that $\mathrm{Div}_{q,q}(\mathbb{C})$ is homotopy equivalent to K_q up to dimension q . The theorem now follows immediately from (3.5).

<u>Corollary (3.8).</u> $\mathrm{Rat}(p,q)$ <u>has the same homotopy type as</u> $\Omega^2 S^2$ <u>up to dimension</u> $\min(p,q)$. <u>In particular</u> $\pi_1(\mathrm{Rat}(p,q)) \cong \mathbb{Z}$ <u>for all</u> $p,q \neq 0$.

<u>Proof.</u> From (3.5) and (3.7) we only have to show $\pi_1(\mathrm{Rat}(p,q)) \cong \mathbb{Z}$ for $p,q \neq 0$. But in this case we have that

$$\pi_1(\mathrm{Rat}(p,q)) \cong \pi_1(\Omega^2 S^2) \cong \pi_3(S^2) \cong \mathbb{Z}$$

(see Hu [70], page 329).

Q.E.D.

Remarks (3.9).

(i) From (3.2) we have that $\mathrm{Rat}(n,0) \cong \mathrm{Rat}(0,n) \cong \mathbb{C}^n$, i.e. $\mathrm{Rat}(n,0)$ is simply connected so that $\pi_1(\mathrm{Rat}(n,0)) \cong \{1\}$. This result was also gotten by Brockett [12] using specific coordinate neighborhoods.

(ii) Brockett [12] also shows that

$$\mathrm{Rat}(n-1,1) \cong \mathrm{Rat}(1,n-1) \cong S' \times \mathbb{R}^{2n-1} .$$

This of course checks with (3.8).

(iii) The homotopy theoretic description of $\mathrm{Rat}(n)$ given above while very important is certainly not the definitive answer. Indeed the questions from identification theory which motivated this study demand explicit coordinate neighborhoods on $\mathrm{Rat}(n)$, i.e. a description of $\mathrm{Rat}(n)$ up to homeomorphism. Of course by (3.3) we have reduced this problem to the study of the components $\mathrm{Rat}(q,q)$. But here there is still a lot of work to be done.

§4. Partial Realizations (Again).

We would like now to discuss some geometric aspects of the partial realization problem that we sketched in Part VI, Section 4 of these notes and again motivated by identification theory to try to give some specific topological structure to the set of minimal realizations. We shall compare two approaches: one due to Brockett [14] and the other to Kalman [81]. We begin with the approach from [81]:

Continued Fractions (4.1). Recall in Part VI, Section 4 that to any infinite sequence $\{a_i\}_{i\geq 1}$ of real numbers, or equivalently to any formal Laurent series $f = \sum_{i\geq 1} a_i z^{-i}$, we associated a set of invariants called jumps corresponding to points where a minimal partial realization of the series up to that point increases dimension. Obviously f is rational if and only if there are finitely many jump points, an assumption which we shall make here. Then if $f = g/h$, $g,h \in \mathbb{R}[z]$, g and h having no common factors, h monic, degree $g <$ degree $h = n$, via the Euclidean algorithm one can represent f as a continued fraction:

$$(*) \qquad f = \cfrac{b_1}{f_1 - \cfrac{b_1 b_2}{f_2 - \cdots - \cfrac{b_{t-1}\, b_t}{f_t}}} .$$

Note that since f is strictly proper, all the $b_i \neq 0$, and all the f_i are monic. Moreover the degree f_i = size of the corresponding jump point.

We might add that such continued fraction representations and their relation to Padé approximation are quite classical. See e.g. Perron [13].

Now clearly these continued fractions and in particular the invariants b_i, degree f_i give us an alternate approach to the Hankel matrices when studying the properties of Rat(n). For example, we have:

Proposition (4.2). Notation as in (4.1). The signature of the Hankel associated to f is determined as follows:

(a) The signature of the 0 sequence is 0.

(b) If degree f_i is even, the number of positive and negative eigenvalues each increase by $(\deg f_i)/2$, and so the signature doesn't change.

(c) If degree f_i is odd, the signature increases by 1 for $b_i > 0$, and decreases by 1 for $b_i < 0$.

Proof. This is (4.1) of [81], page 21, and is derived as a corollary of a classical result of Frobenius [34].

Q.E.D.

Remarks (4.3).

(i) If all the b_i have the same sign and all the f_i have degree 1 (the "generic" case classically studied), then $f \in \mathrm{Rat}(n,0)$ if all the $b_i > 0$, and $f \in \mathrm{Rat}(0,n)$ if all the $b_i < 0$.

(ii) It is interesting to compare the b_i regarded as local coordinates on Rat(n) and some of the coordinates given by Brockett in [12] for Rat(n). We shall consider here the case $n = 2$, and the component Rat(1,1) (this will be seen to be sufficiently complicated and sufficiently messy).

In this case we have two possibilities for the continued fraction expansion of an element of Rat(1,1):

(a) $f = \dfrac{b_1}{f_1 - b_1}$ where f_1 is monic of degree 2 and $b_1 \neq 0$.

This is the non-generic case and these functions depend on three parameters: namely b_1 and the coefficients of f_1. Hence the set of all these functions is homeomorphic to $U := \{(a,b,c) \in \mathbb{R}^3 \mid c \neq 0\}$.

(b) $f = \dfrac{b_1}{f_1 - \dfrac{b_1 b_2}{f_2}}$

where $b_1 b_2 < 0$ and the f_i are monic polynomials of degree 1. Clearly the subset of Rat(1,1) consisting of functions of type (b) is homeomorphic to $V := \{(a,b,c,d) \in \mathbb{R}^4 \mid cd < 0\}$ which is four-dimensional. This case is the generic case.

Now we want to indicate how the sets U and V fit together to get the smooth manifold Rat(1,1). One method is to consider Brockett's ([12]) parametrization of rational functions in Rat(1,1):

$$f(z) = \frac{e^{\alpha}[(z+\sigma)\cos\theta + \sin\theta]}{(z+\sigma)^2 + \ln(e^{\beta} + e^{-\tan^2\theta})}$$

where $(\alpha,\beta,\sigma) \in \mathbb{R}^3$, $\theta \in S^1$. From this parametrization it is an elementary and slightly tedious exercise to construct explicit homeomorphisms

$$\mathbb{R}^3 \times (S^1 - \{\frac{\pi}{2}, \frac{3\pi}{2}\}) \xrightarrow{\sim} V$$

$$\mathbb{R}^3 \times \{\frac{\pi}{2}, \frac{3\pi}{2}\} \xrightarrow{\sim} U$$

which allows us to paste U and V together.

Definitions (4.4).

(i) Set

$$R(r,n) := \{(a_1,\dots,a_r) \in \mathbb{R}^r \mid \text{the partial sequence } a_1,\dots,a_r \text{ admits a minimal realization of dimension } n\}.$$

We give $R(r,n)$ the topology induced from $\mathbb{R}^r$.

(ii) We let $\mathcal{H}(n,k)$ denote the set of real $n \times n$ Hankel matrices of rank k. We regard $\mathcal{H}(n,k)$ as a subset of $\mathbb{R}^{2n}$ with the induced topology.

(iii) Set

$$\mathcal{H}(n,k)_1 := \{H \in \mathcal{H}(n,k) \mid \text{the first } k \times k \text{ minor is singular}\} \;;$$

$$\mathcal{H}(n,k)_0 := \{H \in \mathcal{H}(n,k)_1 \mid \text{the } (n,n) \text{ entry is } 0\} \;.$$

We can now state the following theorem from [14] about the topology of the partial realization problem:

Theorem (4.5).

(i) $\mathcal{H}(n,k)$ is a real analytic manifold of dimension $2k$ for $k < n$, $2k-1$ for $k = n$. Moreover $\mathcal{H}(n,k)$ has $k+1$ connected components.

(ii) $\mathcal{H}(n,k)_1$ is a real analytic manifold of dimension $2k-1$ for $0 < k < n$ of which $\mathcal{H}(n,k)_0$ is a $2k-2$ dimensional submanifold.

(iii) $R(r,n)$ is homeomorphic to $\mathcal{H}(n,n) \times \mathbb{R}$ for $2n \leq r$, $\mathcal{H}(\frac{r+1}{2}, r-n+1)_1$ for $2n > r$ and r odd, $\mathcal{H}(\frac{r+2}{2}, r-n+1)_0$ for $2n > r$ and r even. In particular, $R(r,n)$ may be given a natural structure as an analytic manifold.

Proof. Since we won't need this result or the techniques used in proving it in the sequel we refer the reader to [14]. We only remark here that (iii) is essentially a geometric restatement of certain known results in partial realization theory. For example the result that for $2n > r$ and r odd that $R(r,n)$ is homeomorphic to $\mathcal{H}(\frac{r+1}{2}, r-n+1)_1$ follows immediately from Kalman [82], page 331.

Q.E.D.

PART VIII. FEEDBACK AND STABILIZATION OF SYSTEMS WITH PARAMETER UNCERTAINTY

We want to conclude these lecture notes with a topic which was perhaps the main motivation for introducing the whole subject of control theory: stabilization and especially stabilization through feedback.

As we will see in Section 1, the idea of stability theory is to see whether a given equilibrium position is stable under small perturbations. This leads then to "controlled" returns to equilibria which leads naturally in turn to feedback. (For good classical discussions of this see Horowitz [66], Jacobs [74], and Luenberger [93].) These concepts were introduced already by James Clerk Maxwell [100] who was interested in the control of the Watt steam engine.

We shall of course consider the modern problems and research on the subject in Sections 2, 3, and 4. However in order to give the pure mathematician a flavor of problems in actual system design, in Section 6 we will give an optimal synthesis procedure for solving a certain (but the most important) case of a problem known as the "blending problem" in the engineering literature.

§1. Classical Stability Theory

In order to keep these notes as self-contained as possible, we will briefly go over some of the basic facts on classical stability theory. The basic reference for this is Gantmacher Vol. II [39]. Another good reference is Marden [97]. For treatments from a control theoretic point of view there are Casti [22], Jacobs [74], and Luenberger [93]. We will concentrate here on the continuous time case, even though of course there is an analogous theory for the discrete time case.

Definition (1.1). Let $\dot{x}(t) = f(x(t))$ be a free (i.e. without controls) time-invariant dynamical system of dimension n (i.e. the state space is k^n, $k = \mathbb{R}$ or $\mathbb{C}$) with f continuous. An equilibrium point for this system is a vector $x_0 \in k^n$ such that $f(x_0) = 0$.

Remark (1.2). Note this definition means that if once the state vector is equal to x_0, then it remains equal to x_0 always. This formulation applies to any free dynamical system not only continuous, time-invariant ones.

We now have the following fundamental definitions:

Definitions (1.3).

(i) In the state space k^n ($k = \mathbb{R}$ or $\mathbb{C}$) of the free time-invariant dynamical system $\dot{x}(t) = f(x(t))$, we let $B_r(x_0)$ be the Euclidean open ball of radius

$r > 0$ around x_0 , x_0 an equilibrium point for the system.

(ii) x_0 is stable if there exists an $\varepsilon' > 0$ such that for every $\varepsilon < \varepsilon'$, there exists a $\delta < \varepsilon$ such that if $x(0) \in B_\delta(x_0)$, then $x(t) \in B_\varepsilon(x_0)$ for all $t > 0$.

(iii) x_0 is asymptotically stable if it is stable and there exists $\varepsilon'' > 0$ such that if $x(t_0) \in B_{\varepsilon''}(x_0)$, then $x(t)$ tends to x_0 as t goes to infinity.

(iv) x_0 is unstable if it is not stable.

Remark (1.4). Let $\dot{x}(t) = Fx(t)$ be a free continuous linear time-invariant dynamical system of dimension n . We will assume that $x(0) \neq 0$. Now clearly $x = 0 \in k^n$ is an equilibrium point, and from now on we will always assume that stability is always taken with respect to 0 .

We now have the following elementary but crucial result whose proof will be left as a simple exercise (or see any of the above references):

Proposition (1.5). Notation as in (1.4). Then the system $\dot{x} = Fx$ is stable if and only if all the eigenvalues of F lie in the closed left half plane $\bar{H} = \{z \in \mathbb{C} \mid \operatorname{Re} z \leq 0\}$ and is asymptotically stable if and only if all the eigenvalues lie in the open left half plane $H = \{z \in \mathbb{C} \mid \operatorname{Re} z < 0\}$.

Routh Criterion (1.6).

(i) We say a polynomial $f \in \mathbb{C}[z]$ is Hurwitz if all its zeroes lie in $\bar{H}$, and strict Hurwitz if all its zeroes lie in H . Then (1.5) means that $\dot{x} = Fx$ is stable if and only if the characteristic polynomial $\det(zI - F)$ is Hurwitz and is asymptotically stable if and only if $\det(zI - F)$ is strict Hurwitz.

(ii) (1.5) has led to a huge literature on the geometry of the zeroes of real and complex polynomials and in particular when such polynomials are Hurwitz. Probably the most famous result along these lines is due to Routh [123] and is discussed in some detail in Gantmacher Vol. II [39] and Marden [97]. Briefly, suppose that $f(z) = z^n + a_1 z^{n-1} + \dots + a_n$ is a real polynomial which for simplicity we assume has no pure imaginary roots. We want to count the number of roots of $f(z)$ in the open left half plane $\{z \in \mathbb{C} \mid \operatorname{Re} z < 0\}$. From standard theorems in complex function theory, to compute the number of roots of any polynomial in any domain, one computes the change in the argument along the boundary of that domain. Now Routh showed that regarding our polynomial $f(z)$, if $r_1 :=$ number of roots of $f(z)$ in $\{z \mid \operatorname{Re} z < 0\}$, and $r_2 :=$ number of roots of $f(z)$ in $\{z \mid \operatorname{Re} z > 0\}$, then the Cauchy index (Part VII, Section 1) of the rational function

$$g(z) := \frac{a_1 z^{n-1} - a_3 z^{n-3} + a_5 z^{n-5} - \dots}{z^n - a_2 z^{n-2} + a_4 z^{n-4} - \dots}$$

is exactly equal to $r_1 - r_2$.

(iii) Recall from Part VII (1.5), that the signature of the Hankel matrix associated to $g(z)$ is exactly the Cauchy index. Looking then at (4.2) of Part VII, if we take the continued fraction associated to $g(z)$, we get the result of Kalman [81] that $g(z)$ is strict Hurwitz if and only if all the b_i's are > 0. (In this case the associated Hankel has signature n of course.)

(iv) In the case of a linear continuous time-invariant dynamical system $\dot{x} = Fx + Gu$, $y = Hx$, we also have a notion of stability. Namely, take the transfer function $T(z) = H(zI - F)^{-1}G$. Clearly the poles of $T(z)$ are eigenvalues of F, and hence stability for such a system means that $T(z)$ is holomorphic in $\{z \mid \text{Re } z > 0\}$ while asymptotic stability means that $T(z)$ is holomorphic in $\{z \mid \text{Re } z \geq 0\}$.

(v) In the case of a free linear time-invariant discrete dynamical system $x(t+1) = Fx(t)$, one can show that this system is stable if and only if all the eigenvalues of F lie in $\{z \in \mathbb{C} \mid |z| \leq 1\}$, and is asymptotically stable if all the eigenvalues lie in $\{z \in \mathbb{C} \mid |z| < 1\}$. See e.g. Luenberger [93] for details.

We conclude this section with a sketch of the Lyapunov theory of stability:

Lyapunov Functions (1.7). The idea of Lyapunov [95] (the so-called "second method", for more modern accounts see [22], [93]) is that an equilibrium point of a free time-invariant dynamical system of dimension n, $\dot{x}(t) = f(x(t))$ over $\mathbb{R}$ with F continuous is asymptotically stable if the trajectories of a process starting near the equilibrium point move so as to minimize a certain "energy" function.

More precisely, on the state space $\mathbb{R}^n$ of $\dot{x}(t) = f(x(t))$, given an equilibrium point x_0 and an open neighborhood U of x_0, a Lyapunov fuction V is a continuous function defined on U with continuous first partial derivatives such that:

(a) V has a unique minimum at x_0.

(b) The function $\frac{dV}{dt}(x(t)) = \frac{\partial V}{\partial x_1} f_1(x(t)) + \ldots + \frac{\partial V}{\partial x_n} f_n(x(t))$ (we are using the chain rule and the system equations $\dot{x}(t) = f(x(t))$; the f_i are the component functions of f) is ≤ 0 for all $x \in U$.

The fundamental theorem (again which we state without proof; for a nice modern proof see [93]) is:

Theorem (1.8). *If there exists a Lyapunov function V, then x_0 is stable. If $\frac{dV}{dt} < 0$, then x_0 is asymptotically stable.*

Remark (1.9). In general it may not be clear how to construct a Lyapunov function for a given system even supposing that one exists. In the linear case however, one has a natural candidate. Indeed let $\dot{x}(t) = Fx(t)$ be as in (1.4), where we take the state space to be $\mathbb{R}^n$ now. Again we consider stability relative to $0 \in \mathbb{R}^n$. Let Q be a symmetric positive definite $n \times n$ real matrix and let

$V(x) := (x,Qx)$ be the corresponding positive definite quadratic form (where (y_1,y_2) := the ordinary inner product of $y_1,y_2 \in \mathbb{R}^n$). Then a simple calculation shows that

$$\frac{dV}{dt}(x(t)) = (x,(F^tQ + QF)x)$$

where F^t denotes the transpose of F. Thus by (1.8) we make the requirement that $(x,(F^tQ+QF)x) < 0$. In other words, we have that $0 \in \mathbb{R}^n$ is an asymptotically stable equilibrium point if we can find a positive definite symmetric Q such that $F^tQ + QF$ is negative definite. Similarly if $F^tQ + QF$ is negative semidefinite, then V is a Lyapunov function and 0 is stable.

§2. Kronecker Indices and State Feedback

In this section we will be concerned with some classical and modern problems about state feedbacks and the associated Kronecker indices both over fields and general Noetherian integral domains.

State Feedback Group (2.1).

(i) Let k be an arbitrary field. Then the state feedback group $\mathcal{F}(n,m)$ is the subgroup of $GL(n+m,k)$ consisting of matrices of the form

$$\begin{bmatrix} A & 0 \\ L & B \end{bmatrix}$$

where $A \in GL(n,k)$, $B \in GL(m,k)$, and $L \in \mathrm{Hom}_k(k^n,k^m)$.

(ii) $\mathcal{F}(n,m)$ acts on the space $k^{n^2+nm} \cong M_{n,n}(k) \times M_{n,m}(k)$ by $(F,G) \longmapsto (AFA^{-1} + AGLA^{-1}, AGB^{-1})$.

(iii) Jt is easy to show that if $V_{n,m} := \{(F,G) \text{ completely reachable}\}$, then $\mathcal{F}(n,m) \times V_{n,m} \to V_{n,m}$.

(iv) $\mathcal{F}(n,m)$ may also be thought of as the semidirect product of the three groups $GL(n,k)$, $GL(m,k)$, $\mathrm{Hom}(k^n,k^m)$, where $GL(n,k)$ acts by change of basis on the state space k^n, $GL(m,k)$ acts by change of base on the space of input values k^m, and $\mathrm{Hom}_k(k^n,k^m)$ is the state feedback part $(F,G) \mapsto (F+GL,G)$ for $L \in \mathrm{Hom}_k(k^n,k^m)$. The latter derives its name from the fact that if we regard (F,G) as the control part of a system $\Sigma = (F,G,H)$, then we have a diagram

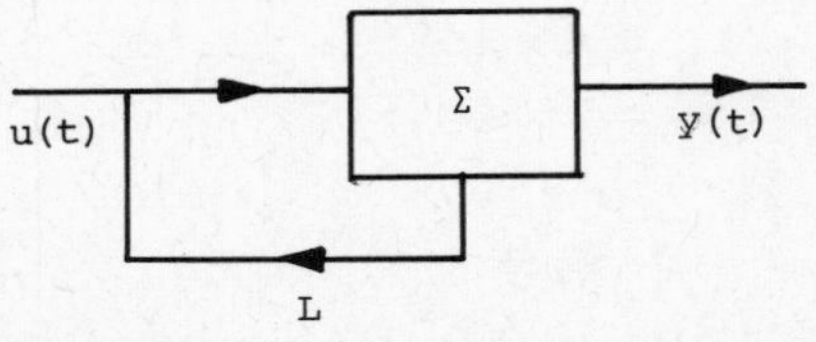

representing the action of L (i.e. we have $u(t) = Lx(t)$ for $u(t)$ the input vector and $x(t)$ the state vector).

(v) Recall our discussion in Part V, §2, about the Kronecker indices and the flat control canonical form. Then it is an easy exercise (see e.g. Kalman [78]) to show that via the action of the state feedback part of $\mathcal{F}(n,m)$, $(F,G) \to (F+GL,G)$ for (F,G) completely reachable, given F_* as in Part V (2.2) one can transform all the entries denoted by α_{ijk} to 0, and similarly via the $GL(m,k)$ part of $\mathcal{F}(n,m)$, $(F,G) \mapsto (F,Gg^{-1})$ for $g \in GL(m,k)$, one can transform all the α_{ijk} entries of G_* to 0.

We thus have the following theorem due to Brunovsky [15]:

<u>Theorem (2.3).</u> The Kronecker indices form a complete set of arithmetic invariants for the elements of $V_{n,m}$ under the state feedback group. Let $\kappa_{i_1}, \dots \kappa_{i_t}$ be these non-zero Kronecker indices $(i_1 \leq \dots \leq i_t)$ associated to $(F,G) \in V_{n,m}$, where $t = \operatorname{rank} G$. Then we may put (F,G) in the following "Brunovsky canonical form":

$$F_b := \begin{bmatrix} A_{\kappa_{i_1}} & 0 & \dots & 0 \\ 0 & A_{\kappa_{i_2}} & & 0 \\ \vdots & \vdots & & \vdots \\ 0 & 0 & \dots & A_{\kappa_{i_t}} \end{bmatrix}$$

$$G_b := \begin{bmatrix} \tilde{e}_{\kappa_{i_1}} & 0 & \dots 0 & \dots 0 \\ 0 & \tilde{e}_{\kappa_{i_2}} & \dots 0 & \dots 0 \\ \vdots & & \ddots & \\ 0 & 0 & \dots \tilde{e}_{\kappa_{i_t}} & \dots 0 \end{bmatrix}$$

where the $A_{\kappa_{i_j}}$ are $\kappa_{i_j} \times \kappa_{i_j}$ matrices defined by

$$A_{\kappa_{i_j}} := \begin{bmatrix} 0 & 1 & 0 & \dots & 0 \\ 0 & 0 & 1 & \dots & 0 \\ \vdots & & & & \vdots \\ 0 & 0 & 0 & \dots & 0 \\ 0 & 0 & 0 & \dots & 0 \end{bmatrix}$$

for $j = 1,\dots,t$, and the $\tilde{e}_{\kappa_{i_j}}$ are $\kappa_{i_j} \times 1$

$$\tilde{e}_{\kappa_{i_j}} := \begin{bmatrix} 0 \\ 0 \\ \vdots \\ 0 \\ 1 \end{bmatrix}$$

for $j = 1,\ldots,t$. Finally via the action of $GL(m,k)$ on (F,G) , we can always arrange the Kronecker indices descending order $\kappa_1 \geq \ldots \geq \kappa_m$. Thus in particular on each orbit of $\mathcal{F}(n,m)$ acting on $V_{n,m}$ there is exactly one pair of matrices with Kronecker indices in descending order and in Brunovsky canonical form.

Proof. It is immediate from (2.1)(v) that we can bring matrices into the Brunovsky canonical form. The fact that we can arrange the Kronecker indices in descending order via $GL(m,k)$ we leave as an easy exercise.

Q.E.D.

Remark (2.4). The content of (2.3) is that the orbits of $\mathcal{F}(n,m)$ acting on $V_{n,m}$ are in 1-1 correspondence with partitions of n , $n = \kappa_1 + \ldots + \kappa_m$, $\kappa_1 \geq \kappa_2 \geq \ldots \geq \kappa_m$. Given this, we will usually assume that the Kronecker indices of a given pair $(F,G) \in V_{n,m}$ are in descending order.

We now come to some nice work of Hermann-Martin [98] relating the Kronecker indices and state feedback to the Grassmannian (Part I,§6). We begin with the following:

Definition (2.5). Let $T(z)$ be a $p \times m$ strictly proper rational transfer function as in Part VI, §3. Let $T(z) = N(z)D(z)^{-1}$ be a right coprime factorization and recall that the factors are unique up to a right unimodular factor. We now assume that $T(z)$ is defined over an algebraically closed field k . Then one gets a morphism $\varphi_T : \mathbb{P}^1_k \to Gr(m,m+p)$ by setting

$$\varphi_T(z) := \{(D(z)u, N(z)u) \mid u \in k^p\}$$

for $z \neq \infty$, and

$$\varphi_T(\infty) := \{(u,0) \mid u \in k^p\} .$$

This Hermann-Martin map φ_T has been quite useful in interpreting system-theoretic data geometrically. For example we have:

Proposition (2.6). Let $T(z) = N(z)D(z)^{-1}$ be as in (2.5). Then the MacMillan degree of $T(z)$ (i.e. the dimension of a canonical realization of $T(z)$) is equal to the degree of the curve $\varphi_T(\mathbb{P}^1) \hookrightarrow \mathbb{P}^{\binom{m+p}{m}-1}$ via the Plucker embedding of $Gr(m,m+p)$.

Proof. We only give a sketch here. For complete details see [98] pages 745-748. The main ingredient is that the MacMillan degree of $T(z) = \deg\det D(z)$, a fact whose proof may be found in Rosenbrock [118] as well as [98]. Now by conservation of number ([45]) we can always assume that $\det D(z)$ has distinct roots. Then it is a triviality to check that the intersection number of $\varphi_T(\mathbb{P}^1)$ with the hyperplane at ∞ of $\mathbb{P}(\Lambda^m k^{m+p}) \cong \mathbb{P}^{\binom{m+p}{p}-1}$ (see Part I, §6, for the explicit form of the Plücker embedding) is exactly $\deg\det D(z)$, from which the required result follows.

Q.E.D.

We can now construct an important bundle in system theory again due to Hermann-Martin [98] as well as its relativization due to Byrnes [17]:

Hermann-Martin Bundle (2.7).

(i) We set $R := \mathbb{C}[X_1,\ldots,X_r]$, the polynomial ring in r variables. We let $\Sigma = (R^n,F,G,H)$ be a completely reachable system over R with input space R^m , and output space R^p .

(ii) Consider the homogenized pencil of matrices $\alpha(w,z) := [wF - zI, zG]$. This determines for each fixed $(w,z) \in \mathbb{C}^2 - \{(0)\}$ an R-linear map $R^n \oplus R^m \to R^n$ defined by $\alpha(w,z)(x,u) := wFx - zx + wGu$.

(iii) We let $\mathbb{A}^t$ be affine t-space over $\mathbb{C}$, and $\mathbb{P}^t$ projective t-space over $\mathbb{C}$ for $t \in \mathbb{N}$. Now the maximal ideals of R are in 1-1 correspondence with the points of $\mathbb{A}^r$, and let us denote the maximal ideal corresponding to $p \in \mathbb{A}^r$ by m_p . Then for each $(w,z) \in \mathbb{A}^2 - \{(0)\}$ and for each $p \in \mathbb{A}^r$ we get a linear map $\alpha(p,w,z) : \mathbb{C}^n \oplus \mathbb{C}^m \to \mathbb{C}^n$ by taking $\alpha(w,z) : R^n \oplus R^m \to R^n$ and reducing modulo m_p . This induces a morphism $\alpha' : \mathbb{A}^r \times \mathbb{A}^2 - \{(0)\} \to Gr(m,m+n)$ by defining $\alpha'(p,w,z) := \ker \alpha(p,w,z)$, which by homogeneity desceds to define a morphism $\tilde{\alpha} : \mathbb{A}^r \times \mathbb{P}^1 \to Gr(m,m+n)$.

(iv) Given our (F,G) completely reachable over R , the generalized Hermann-Martin bundle $\mathcal{V}_{(F,G)}$ (see [17]) is defined to be the pull-back of the universal bundle over $Gr(m,m+n)$ (see Part IV (5.1)) to $\mathbb{A}^r \times \mathbb{P}^1$ via $\tilde{\alpha}$.

(v) If $r=0$ (so that $\mathbb{A}^r$ is a point) we get a bundle over $\mathbb{P}^1$ in this way. Now given a $p \times m$ strictly proper rational transfer function $T(z)$, we have the Hermann-Martin map $\varphi_T : \mathbb{P}^1 \to Gr(m,m+p)$ as in (2.5) and we thus get a bundle over $\mathbb{P}^1$ by taking the pull-back of the universal bundle U_m over $Gr(m,m+p)$ by φ_T . Hermann-Martin ([98], p.749) show that if T is realized by a canonical system (F,G,H) then $\varphi_T^* U_m \cong \mathcal{V}_{(F,G)}$. This may be easily seen as follows: Over $s \in \mathbb{P}^1$, $s \neq \infty$, we have that the fiber $\mathcal{V}_{(F,G)}(s) \cong \ker[F - sI, G]$, where $[F - sI, G] : \mathbb{C}^n \oplus \mathbb{C}^m \to \mathbb{C}^n$ by $(x,u) \mapsto Fx - sx + Gu$ ($[F - sI, G]$ is classically called a pencil of matrices). Over $s = \infty$, we have $\mathcal{V}_{(F,G)}(s) = \{(0,u) \mid u \in \mathbb{C}^m\}$. Then define a map from $\mathcal{V}_{(F,G)} \to \varphi_T^* U_m$ by $(x,u) \mapsto (Hx,u)$, which maps the kernel of $[F - sI, G]$ to the graph of T . It then immediately follows from the fact that (F,G,H) is observable, that this map is injective.

Now we want to relate the Kronecker indices with certain invariants associated to the Hermann-Martin bundle. This is done through the following result due to Birkhoff [8] and Grothendieck [46], which we alluded to at the end of Part I, § 5:

Theorem (2.8). Notation as in Part I (5.8) and (5.9). Then any vector bundle $\mathcal{V}$ of rank m over $\mathbb{P}^1$ is isomorphic to a sum of line bundles $\bigoplus_{i=1}^{m} \mathcal{O}_{\mathbb{P}^1}(n_i)$. In particular, the isomorphism classes of rank m , vector bundles on $\mathbb{P}^1$ are in 1-1 correspondence with sets of integers $n_1 \geq \ldots \geq n_m$.

Proof. From the definition of vector bundle of rank m , and from the fact that $\mathbb{P}^1$ admits a coordinate covering $U_0 = \{(x,y) \mid x \neq 0\}$, $U_1 = \{(x,y) \mid y \neq 0\}$ (i.e. we have local coordinates z and $\frac{1}{z}$), the vector bundle must be determined by the glueing data on $U_0 \cap U_1$, i.e. by an element of $GL(m,\mathbb{C}[z, \frac{1}{z}])$. Moreover given two vector bundles represented by matrices M, $N \in GL(m,\mathbb{C}[z, \frac{1}{z}])$, these vector bundles will be isomorphic if and only if there exist matrices $A \in GL(m,\mathbb{C}[z])$ and $B \in GL(m,\mathbb{C}[\frac{1}{z}])$ such that $M = ANB^{-1}$. Now Birkhoff [8] (in connection with his work on the Riemann-Hilbert problem and having nothing to do with vector bundles!) proved that any $M \in GL(m, \mathbb{C}[z, \frac{1}{z}])$ admits a factorization of the form

$$M = A_1 \begin{pmatrix} z^{n_1} & 0 & \cdots & 0 \\ 0 & z^{n_2} & \cdots & 0 \\ \vdots & \vdots & & \vdots \\ 0 & 0 & & z^{n_m} \end{pmatrix} A_2$$

where $A_1 \in GL(m, \mathbb{C}[z])$, and $A_2 \in GL(m, \mathbb{C}[\frac{1}{z}])$ from which we have the theorem.

Q.E.D.

We can thus prove the following result from [98]:

<u>Corollary (2.9).</u> <u>Let</u> $(F_b,G_b) \in V_{n,m}$ <u>be in Brunovsky canonical form</u> (2.3) <u>with Kronecker indices</u> $\kappa_1 \geq \kappa_2 \geq \dots \geq \kappa_m$. <u>Decompose the Hermann-Martin bundle</u> $\mathcal{V}_{(F_b,G_b)} \cong \bigoplus_{i=1}^{m} \mathcal{O}_{\mathbb{P}^1}(n_i)$ <u>as in</u> (2.8) <u>with</u> $n_1 \geq n_2 \geq \dots \geq n_m$. <u>Then</u> $\kappa_i = n_i$ <u>for all</u> $i = 1,\dots,m$.

<u>Proof.</u> We use the notation of (2.3). Now $\mathcal{V}_{(F_b,G_b)}$ is completely determined by the pencil $[F_b-sI,G_b]$, and clearly if we set $G_{bi} := [\tilde{e}_{\kappa_i}\ 0 \dots 0]$, then we have $[F_b-sI,G_b] = \bigoplus_i [A_{\kappa_i}-sI,G_{bi}]$.

Since (A_{κ_i},G_{bi}) is completely reachable, the dimension of the kernel of $[A_{\kappa_i}-sI,G_{bi}]$ is 1, and thus this kernel determines a line bundle. From the form of (A_{κ_i},G_{bi}) it is easy to compute the degree of this line bundle to be κ_i (i.e. it is isomorphic to $\mathcal{O}_{\mathbb{P}^1}(\kappa_i)$). Since the decomposition of (2.8) is unique up to ordering we must have therefore that $\kappa_i = n_i$ as claimed.

Q.E.D.

<u>Feedback Equivalence (2.10).</u>

(i) Let R be as in (2.7) and let (F_1,G_1) and (F_2,G_2) be two pairs of completely reachable matrices over R with the F_i $n \times n$, and the G_i $n \times m$ for $i = 1,2$. Then we say that (F_1,G_2) is <u>feedback equivalent</u> to (F_2,G_2) (denoted by $(F_1,G_1) \sim (F_2,G_2)$) if there exist $A_1 \in GL(n,R)$, $A_2 \in GL(m,R)$, $L \in Hom_R(R^m,R^n)$ such that

$$F_2 = A_1(F_1 + GL)A_1^{-1}$$

$$G_2 = A_1G_1A_2 \ .$$

(ii) Given the associated pencils $[F_1-sI,G_1]$, $[F_2-sI,G_2]$ it is clear that $(F_1,G_1) \sim (F_2,G_2)$ if and only if

$$(*) \quad A_1[F_1-sI,G_1]\begin{bmatrix} A_1^{-1} & 0 \\ LA_1^{-1} & A_2 \end{bmatrix} = [F_2-sI,G_2] .$$

From this it is immediate that if $(F_1,G_1) \sim (F_2,G_2)$, then $\mathcal{V}_{(F_1,G_1)} \cong \mathcal{V}_{(F_2,G_2)}$. Two pencils satisfying (*) are said to be <u>strictly equivalent</u>. (See Gantmacher, Vol. II [39].)

(iii) If $R = \mathbb{C}$, then it is immediate from (2.3), (2.8) and (2.9) that conversely if $\mathcal{V}_{(F_1,G_1)} \cong \mathcal{V}_{(F_2,G_2)}$, then $(F_1,G_1) \sim (F_2,G_2)$. Indeed from (2.3) completely reachable pairs are classified up to feedback equivalence by sets of integers $\kappa_1 \geq \ldots \geq \kappa_m$, by (2.8) rank m vector bundles are classified up to isomorphism by sets of integers $n_1 \geq \ldots \geq n_m$, and by (2.9) for $(F,G) \in V_{n,m}$ and $\mathcal{V}_{(F,G)}$ these sets are the same.

(iv) The result of (iii) does generalize to arbitrary polynomial rings R, but the proof is much more difficult and in point of fact is one of the key results of C. Byrnes in [17]. The problem is as follows: Suppose we try to directly mimic the proof of (iii) via the Kronecker indices. So let (F,G) be a completely reachable pair over the polynomial ring $R = \mathbb{C}[X_1,\ldots,X_r]$ and let $\mathcal{V}_{(F,G)}$ be the Hermann-Martin bundle over $\mathbb{A}^r \times \mathbb{P}^1$. Then for each $p \in \mathbb{A}^r$ we have that

$$\mathcal{V}_{(F,G)}\Big|_{p \times \mathbb{P}^1} \cong \bigoplus_{i=1}^{m} \mathcal{O}(\kappa_i(p))$$

where the $\kappa_i(p)$ are the Kronecker indices of the pair (F,G) locally at p (this is easy to see from our above discussion).

Now while these local Kronecker indices are constant on a dense Zariski open subset of $\mathbb{A}^r$ (and in point of fact if we regard (F,G) as defined over K the quotient field of R and take the corresponding Kronecker indices κ_i, these will be equal almost everywhere to the local Kronecker indices; see [17], page 1366), their values may change on a closed subvariety of $\mathbb{A}^r$. Because of this we may have pairs (F_i,G_i), $i = 1,2$ with the same local Kronecker indices but which are not feedback equivalent (for an explicit example see [17], page 1366, Example (5.1)).

Thus we can state now the following nice theorem from [17] (page 1369):

<u>Theorem (2.11).</u> <u>Notation as in</u> (2.10). <u>Then</u> $(F_1,G_1) \sim (F_2,G_2)$ <u>if and only if</u> $\mathcal{V}_{(F_1,G_1)} \cong \mathcal{V}_{(F_2,G_2)}$.

<u>Proof.</u> We give only a brief sketch here. For full details see the above reference.

From (2.10) we need only show that $\mathcal{V}_{(F_1,G_1)} \cong \mathcal{V}_{(F_2,G_2)}$ implies that $(F_1,G_1) \sim (F_2,G_2)$. But if M_i: = R-module of global sections of $\mathcal{V}_{(F_i,G_i)}$,

$i = 1,2$, then it is easy to show that M_i is finitely generated projective R-module of rank $n+m$ and hence by the Quillen-Suslin theorem [116], free. The isomorphism of the vector bundles of course induces an isomorphism of R-modules $\varphi\colon M_1 \to M_2$. Now by construction the module of global sections of $\mathcal{V}_{(F_i,G_i)}|\mathbb{A}^r\times\{\infty\}$ over $\mathbb{A}^r \times \{\infty\}$ is R^m the input space, and hence we get a surjection of R-modules $M_i \to R^m$ induced by $\mathbb{A}^r \times \mathbb{P}^1 \to \mathbb{A}^r \times \{\infty\}$. Since the M_i are free we get isomorphisms $M_i \cong R^n \oplus R^m$, and φ induces an isomorphism $\tilde{\varphi}\colon R^n \oplus R^m \to R^n \oplus R^m$ which takes the input space R^m to R^m (since $\tilde{\varphi}$ is induced by a bundle isomorphism it takes the global sections of $\mathcal{V}_{(F_1,G_1)}|\mathbb{A}^r\times\{\infty\}$ to the global sections of $\mathcal{V}_{(F_2,G_2)}|\mathbb{A}^r\times\{\infty\}$) Thus $\tilde{\varphi}$ has the form of the block 2×2 matrix in Equation (*) of (2.10) (ii) from which the result easily follows.

Q.E.D.

<u>Corollary (2.12).</u> <u>Let</u> (F,G) <u>be a completely reachable pair of matrices over the polynomial ring</u> R <u>(with</u> F $n\times n$, G $n\times m$). <u>Then if the local Kronecker indices ((2.10) (iv)) of</u> (F,G) <u>are constant,</u> (F,G) <u>is feedback equivalent to a completely reachable pair defined over</u> $\mathbb{C}$.

<u>Proof.</u> We will give two proofs here. The first is the original proof due to Byrnes [17], and the second one is due to Hazewinkel [56].

The first proof goes as follows: From the remarks of (2.10)(iv) since the $\kappa_i(p)$ are constant as functions of $p \in \mathbb{A}^r$ we have that

$$\mathcal{V}_{(F,G)}\Big|_{p\times\mathbb{P}^1} \cong \bigoplus_{i=1}^{m} \mathcal{O}_{\mathbb{P}^1}(\kappa_i)$$

for each $p \in \mathbb{A}^r$. But then from Theorem 7, page 199 of Hanna [51] and the Quillen-Suslin theorem, this implies that

$$\mathcal{V}_{(F,G)} \cong \bigoplus_{i=1}^{m} \pi_2^* \mathcal{O}_{\mathbb{P}^1}(\kappa_i)$$

where $\pi_2\colon \mathbb{A}^r \times \mathbb{P}^1 \to \mathbb{P}^1$ is the projection, and thus by (2.11) we are done.

The second proof makes use of ideas from Part V. We regard (F,G) as an algebraic family of matrix pairs parametrized by $\mathbb{A}^r$ in the now standard way, and for each $p \in \mathbb{A}^r$ by hypothesis $\kappa_i(p) = \kappa_i$. Now we have seen that the control canonical form of Part V, Section 2 is a holomorphic canonical form and from this it is easy to see that it must be an algebraic canonical form. The content of (2.3) of Part V is that the reduction of completely reachable pairs to control canonical form is holomorphic in the parameters as long as we fix the Kronecker indices as well as the Kronecker nice selection. Thus suppose for our family of pairs of matrices (F,G) with constant Kronecker indices we can find a $g \in GL(m,R)$ such that for (F,Gg) the Kronecker nice selection is also constant. Then by the above via the action of $GL(n,R)$ we can bring (F,Gg) to control canonical form, and then it is easy to see (from the proof of (2.3) of this section) that via the action of $\mathrm{Hom}_R(R^n,R^m)$ and $GL(m,R)$ we can kill the corresponding α_{ijk}'s and hence bring

(F,Gg) to Brunovsky canonical form.

Thus we must show that we can find a $g \in GL(m,R)$ such that for (F,Gg) the Kronecker nice selection is constant. First set $M_i := $ R-module generated by the columns of $G, FG, \ldots, F^iG$ for each $i = 0,\ldots,n-1$. We claim that since the local Kronecker indices are constant, M_i/M_{i-1} is a locally free R-module for each $i = 0,\ldots,n-1$ where $M_{-1} := 0$. Indeed if we let $a_i(p) :=$ dimension of the span of $G(p)$, $F(p)G(p),\ldots,F^i(p)G(p)$ for each $p \in \mathbb{A}^r$ and $i = 0,\ldots,n-1$, then the local Kronecker indices completely determine (and are determined) by the $a_i(p)$. Hence the constancy of the Kronecker indices implies that the $a_i(p)$ are constant as well from which the claim follows immediately. But since R is a polynomial ring, again by the Quillen-Suslin theorem each M_i/M_{i-1} is a free R-module for $i = 0,\ldots,n-1$. Then by considering the action of $GL(m,R)$ step by step on $G,FG,\ldots,F^{n-1}G$, from the above it is an easy exercise in linear algebra to show that we can find a $g \in GL(m,R)$ such that (F,Gg) has constant Kronecker nice selection. We leave the details to the reader or see Hazewinkel [56], pages 42-43.

Q.E.D.

Orbit Space Under State Feedback (2.13). It is also of geometric and system-theoretic interest to compute explicitly the geometric structure of the orbits of the space

$$S(\kappa_1,\ldots,\kappa_m) := \{(F,G,H) \mid (F,G) \in V_{n,m} \text{ with Kronecker indices } \kappa_1,\ldots,\kappa_m\}$$

under the action of the state feedback group $\mathcal{F}(n,m)$. Indeed Wang-Davison [150] and Brockett [13] have explicitly computed the stabilizer subgroup of a pair of matrices in Brunovsky canonical form under the action of the state feedback group. Let us call this stabilizer subgroup $\mathcal{G}$. Then since we may reduce any completely reachable pair to Brunovsky canonical form with the Kronecker indices arranged in descending order, to compute $S(\kappa_1,\ldots,\kappa_m)/\mathcal{F}(n,m)$, it suffices to compute $M_{p,n}(k)/\mathcal{G}$ (where p = dimension of the output space).

In the generic case with $\kappa_1 = \kappa_2 = \ldots = \kappa_m = n/m$, this was done by Kalman [80]. He computes that the orbit space is isomorphic to the Grassmannian $Gr(m,pn/m)$. Kalman also computes that for $n = 3$, $m = 2$, $p = 2$, $\kappa_1 = 2$, $\kappa_2 = 1$, the orbit space is isomorphic to $\mathbb{P}^1 \times \mathbb{P}^1$.

Actually Kalman's results generalize in a rather straightforward way. We assume that for $S(\kappa_1,\ldots,\kappa_m)$ we have that $\kappa_1 \geq \ldots \geq \kappa_m > 0$ so that G has maximal rank. Then a rather messy calculation shows that if we have Kronecker indices such that

$$\kappa_1 > \kappa_2 > \ldots > \kappa_{s-1} > \kappa_s = \kappa_{s+1} = \ldots = \kappa_{s+i} > \kappa_{s+i+1} \ldots$$

then

$$S(\kappa_1,\ldots,\kappa_m) \cong \mathbb{P}^{p\kappa_1-\lambda_1} \times \ldots \times \mathbb{P}^{p\kappa_{s-1}-\lambda_{s-1}} \times Gr(i+1,p\kappa_s-(\lambda_s-i-1)) \times \ldots$$

where

$$\lambda_i := \sum_{j=i}^{m} (\kappa_i-\kappa_j+1) \quad \text{for } i = 1,\ldots,m .$$

Since we won't need this result in the sequel, we won't give the tedious computation here.

§3. Coefficient and Pole Assignability

This section concerns how we may modify the behavior and in particular the stability properties of a given time-invariant linear dynamical system by using state feedback. More specifically, let $\dot{x}(t) = Fx(t) + Gu(t)$ be a linear time-invariant real or complex system where F is $n \times n$ and G is $n \times m$, $x(0) \neq 0$. Then we have that $x(t) = e^{Ft}x(0) + \int_0^t e^{F(t-s)}Gu(s)ds$. From this representation it is clear that if the free system $\dot{x}(t) = Fx(t)$ is unstable, then for any choice of controls $u(t)$, the controlled system $\dot{x}(t) = Fx(t)+Gu(t)$ will be unstable (i.e. $x(t)$ is unbounded for any choice of $u(t)$). Moreover, it is also clear that if $\dot{x}(t) = Fx(t)$ is asymptotically stable, then $\dot{x}(t) = Fx(t)+Gu(t)$ will also be if $u(t)$ is such that $\lim_{t\to\infty}\int_0^t Gu(s)ds < \infty$ (i.e. $x(t)$ goes to 0 as t goes to ∞ for such control functions $u(t)$).

The above discussion means that in general we cannot modify the stability properties of a system by use of open loop input functions $u(t)$. This leads us to try closed loop or state feedback functions of the form $u(t) = Lx(t)$ (where L is $m \times n$) to try to modify the stability properties. As we have seen in (2.1)(iv) such a transformation L, transforms the matrix pair (F,G) to $(F+GL,G)$ (i.e. since $\dot{x} = Fx+Gu$, if $u = Lx$ we have $\dot{x} = (F+GL)x$); this we have called "state feedback". Modifications of system behavior via state feedback then will be our topic in this section.

We begin with the following standard definitions:

Definitions (3.1).

(i) Let (F,G) be a pair of matrices with coefficients in an arbitrary commutative ring with unity R, F $n \times n$, G $n \times m$. Then (F,G) is said to be pole assignable if for every $s_1,\dots,s_n \in R$, there exists an $m \times n$ matrix L with coefficients in R such that $\det(zI-F-GL) = (z-s_1)\dots(z-s_n)$.

(ii) (F,G) as in (i) is said to be coefficient assignable if given any monic polynomial $\lambda(z) \in R[z]$ of degree n, there exists L $m \times n$ such that $\det(zI-F-GL) = \lambda(z)$.

Remarks (3.2).

(i) The term "pole assignable" comes from the fact that if we regard (F,G) as defining the system $\dot{x} = Fx+Gu$ (we are ignoring the output part here) where we take the ring R to be the field $\mathbb{R}$ or $\mathbb{C}$, then the poles of the transfer function $T(z) = (zI-F)^{-1}G$ are clearly roots of $\det(zI-F)$, and if (F,G) is pole assignable we can therefore assign the poles of $T(z)$ arbitrary values via state feedback transformations. From our discussion in Section 1 (see especially (1.5) and (1.6) (iv)), this means in the pole assignable case we can modify the stability properties of a system as we like through state feedback.

(ii) Over an arbitrary ring R it is clear that coefficient assignability is a stronger property than pole assignability.

(iii) In Theorem (3.5) below we are going to show that over a field, coefficient assignability is equivalent to complete reachability. Over $\mathbb{C}$, the result is referred to as the "pole shifting theorem" for obvious reasons. This theorem has a number of proofs e.g. Langenhop [92] and Wonham [152] give proofs for (F,G) defined over an infinite field, and Kalman [75] over a general field k. The simplest proof of this result over a general field is due to M. Heymann [62] and we would like to give his neat proof here. The proof is based on two lemmas, this second of which (3.4) is known as "Heymann's lemma" :

Lemma (3.3).

(i) Let (F,G) and R be as in (3.1)(i). If (F,G) is coefficient assignable, then (F,G) is completely reachable.

(ii) Suppose G is $n \times 1$. Then if (F,G) is completely reachable, (F,G) is coefficient assignable.

Proof.

(i) Suppose that (F,G) is coefficient assignable. Then clearly for every maximal ideal $m \subset R$, the associated pair of matrices $(F(m),G(m))$ defined over the residue field R/m will be coefficient assignable, and since complete reachability is a local property (Part VI (5.10)(ii)), it suffices to prove that over a field k coefficient assignability implies complete reachability. So we assume now that (F,G) is a coefficient assignable pair defined over k. Now in this case we can use the following argument of Kalman [75], page 64:

Suppose (F,G) is not completely reachable. Then using the fact that the subspace of the state space consisting of the reachable states is F-invariant after change of basis in the state space if necessary, we may assume that F and G have the following forms:

$$F = \begin{bmatrix} F_{11} & F_{12} \\ 0 & F_{22} \end{bmatrix} , \qquad G = \begin{bmatrix} G_1 \\ 0 \end{bmatrix}$$

and (F_{11},G_1) is completely reachable. But $\det(zI-F) = \det(zI-F_{11}) \det(zI-F_{22})$. From this it is immediate that for any L $m \times n$, $\det(zI-F_{22})$ divides $\det(zI-F-GK)$ which means that (F,G) is not coefficient assignable.

(ii) If (F,G) over R is completely reachable with G $n \times 1$, then we can reduce (F,G) to control canonical form (see Part II, Section 4 and Part V, Section 2) via change of basis in the state space R^n:

$$F_* = \begin{bmatrix} 0 & 1 & 0 & \cdots & 0 & 0 \\ 0 & 0 & 1 & \cdots & 0 & 0 \\ \cdot & \cdot & \cdot & \cdots & \cdot & \cdot \\ 0 & 0 & 0 & \cdots & 0 & 1 \\ -\alpha_n & -\alpha_{n-1} & -\alpha_{n-2} & \cdots & -\alpha_2 & -\alpha_1 \end{bmatrix} , \qquad G_* = \begin{bmatrix} 0 \\ \cdot \\ \cdot \\ 0 \\ 1 \end{bmatrix}$$

where $\det(zI-F) = z^n+\alpha_1 z^{n-1} + \ldots + \alpha_n$. Note that the proof for this given in [82], page 43 over a field works word for word over an arbitrary commutative ring with unity. Now suppose that $\lambda(z) = z^n+a_1z^{n-1} + \ldots + a_n$ with $a_i \in R$, $i = 1,\ldots,n$. Then we let $L: = [\alpha_n-a_n \ \ldots \ \alpha_1-a_1]$ where we write the linear transformation $L: R^n \to R$ in matrix form with respect to the same basis of R^n which we used to represent (F,G) in control canonical form (F_*,G_*). Immediately we have $\det(zI-F_*-G_*L) = \lambda(z)$.

Q.E.D.

Lemma (3.4). Let (F,G) be a completely reachable matrix pair over a field k with F $n\times n$, G $n\times m$, and let g be any non-zero column vector of G. Then there exists a matrix L such that $(F+GL,g)$ is completely reachable.

Proof. We sketch the proof here and refer the reader for details of the computations to [62], pages 748-749. By permuting the columns of G if necessary we can assume that $g = g_1$ is the first column of G, and that if $\kappa_1,\ldots,\kappa_t$ are the non-zero Kronecker indices $(t = \text{rank } G)$, then the $n\times n$ matrix

$$M: = [g_1 \ Fg_1 \ \ldots \ F^{\kappa_1-1}g_1 \ g_2 \ \ldots \ F^{\kappa_t-1}g_t]$$

is non-singular. Set $\mu_r: = \sum_{j=1}^{r} \kappa_j$ and let $e_r: =$ r-th $m\times 1$ standard basis vector for $r = 1,\ldots,t$. Define an $m\times n$ matrix N with columns c_i $(1 \leq i \leq n)$, by setting $c_{\mu_s}: = e_{s+1}$ for $s = 1,\ldots,t-1$, and $c_i: = 0$ otherwise. Then one may explicitly compute that if we set $L: = -NM^{-1}$, $(F+GL,g)$ is completely reachable.

Q.E.D.

Theorem (3.5). Over a general field k, complete reachability is equivalent to coefficient assignability.

Proof. From (3.3)(i) we need only show complete reachability implies coefficient assignability. Let then (F,G) be completely reachable. We can suppose that the first column g_1 of G is non-zero. Then from (3.4) there exists an L $m\times n$ such that $(F+GL,Ge_1)$ is the first standard basis vector of k^m regarded as a column vector. The result now follows immediately from (3.3)(ii).

Q.E.D.

Remarks (3.6).

(i) Bumby-Sontag [16] have recently produced an example which shows in general that over arbitrary R for $m > 1$, complete reachability does not imply coefficient assignability.

(ii) The most powerful result over a field k about what can be altered by state feedback is due to Rosenbrock [118], page 190. His theorem states that for (F,G) completely reachable, a state feedback transformation of F of the type $F \longrightarrow F+GL$ can arbitrarily alter the invariant factors ψ_i of F subject only to the following conditions on the degrees of the ψ_i: $\deg \psi_1 \geq \kappa_1$, $\deg \psi_1 + \deg \psi_2 \geq \kappa_1+\kappa_2$, ... where the Kronecker indices κ_i and the $\deg \psi_i$ are arranged in decreasing order.

(iii) Finally there is an ordering implied by the preceding theorem of Rosenbrock.

Namely given two sequences of non-negative integers $\kappa = (\kappa_1,\ldots,\kappa_m)$ and $\kappa' = (\kappa_1',\ldots,\kappa_m')$ arranged in decreasing order with $\sum_{i=1}^{m} \kappa_i = \sum_{i=1}^{m} \kappa_i'$, define $\kappa > \kappa'$ if and only if $\sum_{i=1}^{s} \kappa_i \leq \sum_{i=1}^{s} \kappa_i'$ for all $s = 1,\ldots,m$. This ordering turns up in many parts of mathematics and has been termed the "ubiquitous ordering" [18]. For a complete discussion of this see Hazewinkel-Martin [59].

We conclude this section with a result from [17] concerning coefficient assignability over polynomial rings:

Theorem (3.7). *Let (F,G) be a completely reachable pair of matrices defined over the polynomial ring $\mathbb{C}[X_1,\ldots,X_r]$. Then if the local Kronecker indices are constant ((2.10) (iv)), (F,G) is coefficient assignable.*

Proof. By (2.11) (F,G) is feedback equivalent to a completely reachable pair defined over $\mathbb{C}$. Hence the theorem follows from (3.5).

Q.E.D.

§4. Blending and Output Feedback

In this section we report first briefly on some recent work concerning stabilization via gain output feedback and then move on to our main theme in this and the next two sections about stabilization via dynamic output feedback. We first must define the relevant concepts:

Definitions (4.1).

(i) Let $\Sigma = (F,G,H)$ be a constant linear dynamical system defined over a field k with F $n \times n$, G $n \times m$, H $p \times n$. Then an *output feedback transformation* is defined by an $m \times p$ matrix K with coefficients in k which transforms $(F,G,H) \longrightarrow (F+GKH,G,H)$. In other words if $u(t)$ is an input vector and $y(t)$ is an output vector, then $u(t) = Ky(t)$. This may be written schematically as

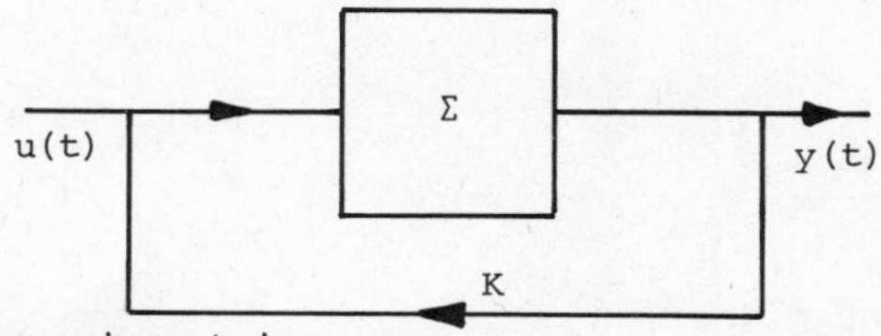

K is called the *gain matrix*.

(ii) The generalization of output feedback which we shall be most interested in here is *dynamic output feedback* in which the output is not only multiplied by a gain matrix K but may be processed by another dynamical system $\tilde{\Sigma}$. That is, one has a diagram

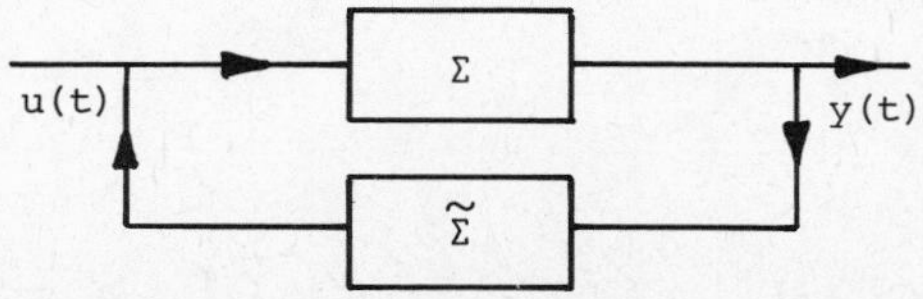

One may show easily (see Chapter 1 of Jacobs [74]) that if the transfer function of Σ is T, and the transfer function of $\tilde{\Sigma}$ is $\tilde{T}$, then the transfer function of the feedback system will be $T/1-T\tilde{T}$.

We now want to quickly sketch some of the results on stabilization and pole placement by gain output feedback of Kimura [84], [85], Hermann-Martin [60], and Byrnes [18].

We begin with [60]:

Proposition (4.2). Let $\Sigma = (F,G,H)$ be as in (4.1)(i) with $k = \mathbb{C}$. Consider the morphism $\varphi_\Sigma: \mathbb{C}^{mp} \to \mathbb{C}^n$ defined by

$$\varphi_\Sigma(K) := \det(zI-F-GKH)$$

where $K \in \mathbb{C}^{mp} \cong M_{p,m}(\mathbb{C})$, and where we identify the monic polynomial $\det(zI-F-GKH)$ with a point in $\mathbb{C}^n$ via its coefficients. Then there exists a Zariski dense open subset U of $\mathbb{C}^{n^2+nm+np}$ such that for all $\Sigma = (F,G,H) \in U$, $\varphi_\Sigma(\mathbb{C}^{mp})$ is a Zariski open dense subset of $\mathbb{C}^n$ when $mp \geq n$.

Proof. For details see [60]. The main idea is to explicitly compute the differential of φ_Σ and to show that under the above hypothesis it is almost everywhere (i.e. on a dense Zariski open subset) maximal rank for a "generic" system Σ (i.e. for all systems lying in a dense Zariski open subset U of $\mathbb{C}^{n^2+nm+np}$). Then by standard results on dominant morphisms (i.e. morphisms with dense images; see [52] or [71]) we have the theorem.

Q.E.D.

Remarks (4.3).

(i) (4.2) means that one may place poles gneerically when $mp \geq n$ over $\mathbb{C}$ via gain output feedback.

(ii) Unfortunately the proof of (4.2) works only over $\mathbb{C}$ and not over $\mathbb{R}$. Over $\mathbb{R}$ using strictly linear algebra Kimura [84] proves the following pole placement theorem (we state his result without proof here; for an account of a nice proof of this due to R. Brockett see [18]):

Theorem (4.4). Let $\varphi_\Sigma: \mathbb{R}^{mp} \to \mathbb{R}^n$ be the analogous map over $\mathbb{R}$ as that defined in (4.2). Then for a generic system Σ, φ_Σ is open and dense when $m+p-1 \geq n$.

Remark (4.5). Note the hypothesis $m+p-1 \geq n$ in (4.4) is stronger than $mp \geq n$ in (4.2).

Finally by studying the Hermann-Martin map associated to the transfer function of a linear time invariant dynamical system as well as using results of Chern [24] about rational curves in the Grassmannian, C. Byrnes [18] has derived the following nice theorem again stated without proof:

Theorem (4.6). Let φ_Σ be as in (4.2) or (4.4). Then generically φ_Σ is proper (i.e. the inverse image of a compact set is compact) when $mp \leq n$. Consequently over $\mathbb{C}$, $\varphi_\Sigma(\mathbb{C}^{mp})$ is a closed subvariety of $\mathbb{C}^n$. Moreover over $\mathbb{R}$, φ_Σ extends to a map of spheres $S^{mp} \to S^n$ and $\varphi(\mathbb{R}^{mp})$ is a Euclidean closed subset of $\mathbb{R}^n$.

Remarks (4.7).

(i) A problem with the above results from an applied standpoint (and apparently inherent in the use of gain output feedback in pole placement) is that in the scalar input/output case $m = p = 1$, we do not get pole placement even generically unless $n = 1$. However the question still remains about the stabilization of systems through dynamic output feedback even in the scalar input/output case.

(ii) This question was treated in a beautiful paper of Youla et al [154] and we will discuss their solution in Section 6. We now would like to state explicitly the problem we shall be concerned with as well as its generalization which is equivalent to a problem known in the engineering literature as the "blending problem" (see Horowitz-Gera [67] and Tannenbaum [147]).

Blending Problem (4.8). Let Σ_1 be a continuous scalar input/output linear time-invariant finite dimensional dynamical system over $\mathbb{R}$. We shall call Σ_1 the plant. Now in [154] the authors consider the problem of stabilizing Σ_1 by means of a dynamic output feedback closed loop as follows: Given Σ_1 we want to find two real linear time-invariant finite dimensional continuous scalar input/output canonical systems Σ_2 and Σ_3 (called the controller and feedback sensor respectively) both asymptotically stable such that the system defined by the closed loop

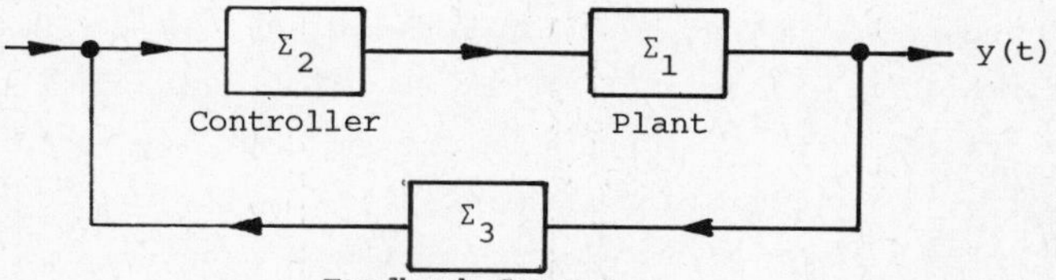

is also asymptotically stable. Such closed loop systems are very desirable from an engineering point of view since instability in the controller tends to result in poor overall system sensitivity to variations in plant parameters.

We now set some notation which we will need here and in Section 6:

$H: =$ open right-half plane

$: = \{z \in \mathbb{C} \mid \operatorname{Re} z > 0\}$

$\bar{H}: =$ closed right-half plane

$: = \{z \in \mathbb{C} \mid \operatorname{Re} z \geq 0\}$

$\tilde{H}: = \bar{H} \cup \{\infty\}$.

Now it is an easy exercise ([154], pages 161-162) that the above stabilization problem reduces to the following problem in complex analysis (the derivation uses the fact that the systems are described by their transfer functions, and one can explicitly write down the transfer function associated to the above closed loop; see (4.1)(ii)): Let $p(z)$ be a real proper rational function (by proper we mean $p(z)$

is finite at ∞ ; for obvious reasons we can even assume that $p(z)$ is strictly proper so that it will be 0 at ∞ , but the proof works the same way in the slightly more general proper case). Then under what conditions can we find a fixed rational real function $g(z)$ such that $1/g(z)$ is proper and holomorphic in $\bar{H}$, and such that $-g(z)+p(z) \neq 0$ for all $z \in \bar{H}$? The solution to this problem will be discussed in Section 6.

The blending problem in all its amorphous forms basically amounts to the question of when the given plant is no longer deterministic but depends on some known parameters and we want to solve an analogous problem to that given above. Mathematically the blending problem can be stated as follows (for a discussion see Tannenbaum [147]): Suppose that we have a family of non-zero proper real rational functions of fixed McMillan degree n such that the numerator and denominator of each member of the family have no common factors, continuously parametrized by a compact set K , $\{p_s(z)\}_{s \in K}$. Then when can we find a fixed real rational function $g(z)$ such that $1/g(z)$ is proper and holomorphic in $\bar{H}$, and such that $-g(z)+p_s(z) \neq 0$ for all $z \in \bar{H}$, $s \in K$?

To simplify our discussion now we will suppress the requirement that the $p_s(z)$ $(s \in K)$ and $g(z)$ have real coefficients and require only that they have complex coefficients, so we have a perhaps more mathematically interesting complex version of the blending problem. (In Section 6 we return to the real case.) Note from standard approximation theorems in complex analysis (see e.g. Hormander [65]) it is enough to find a meromorphic $g(z)$ (not necessarily rational) which satisfies the above requirements i.e. $1/g(z)$ is proper and holomorphic in $\bar{H}$ and $-g(z)+p_s(z) \neq 0$ for all $z \in \bar{H}$, $s \in K$, since one can always approximate such functions by appropriate rational functions.

We now want to give a topological obstruction to a solution of this complex blending problem which will allow us to make contact with the work of Part VII:

Example (4.9). Notation as in (4.8). We now assume however that our family of rational functions is such that each $p_s(z)$ is strictly proper (i.e. $p_s(\infty) = 0$) for every $s \in K$ and of fixed McMillan degree n . Set $p_s(z) = p_{1s}(z)/p_{2s}(z)$ where $p_{is}(z) \in \mathbb{C}[z]$ for each $s \in K$, $i = 1,2$. Then to solve the blending problem we are required in particular to find fixed polynomials $g_i(z) \in \mathbb{C}[z]$, $i=1,2$ such that $p_{1s}(z)g_1(z)+p_{2s}(z)g_2(z) \neq 0$ for all $z \in \bar{H}$, $s \in K$.

Now suppose that there exist $a,b \in H$ $(a \neq b)$ such that $p_{1s}(a) = p_{1s}(b) = 0$ for all $s \in K$ and such that

(i) $p_{2s}(a)$ circles around $0 \in \mathbb{C}$ as s varies in K ;

(ii) $p_{2s}(b)$ is a fixed non-zero constant for all $s \in K$.

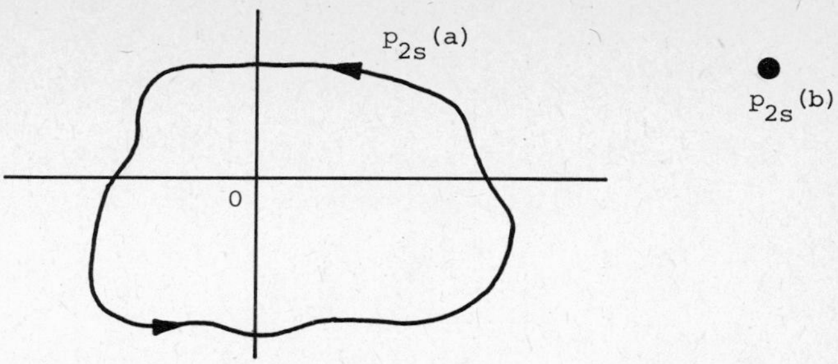

We claim that the blending problem has no solution even if we require g_1 and g_2 to be only continuous. Indeed suppose there existed complex continuous functions g_1, g_2 such that $f_s := g_1 p_{1s} + g_2 p_{2s}$ had no right half plane zeros. Then note that $f_s(a) = g_1(a)p_{1s}(a)+g_2(a)p_{2s}(a) = g_2(a)p_{2s}(a)$ and $g_2(a) \neq 0$ (since otherwise f_s would have a right half-plane zero). Thus at a, the function f_s circles around 0 as s varies in K. Next $f_s(b) = g_1(b)p_{1s}(b)+g_2(b)p_{2s}(b) = g_2(b)p_{2s}(b)$ is a fixed non-zero constant for all $s \in K$ (again as above $g_2(b) \neq 0$). By continuity, then since $a,b \in H$ and the line connecting a and b lies in H, for some point on this line f_s must vanish. (To see this, note that as we move along the line from a to b the closed loop which $f_s(a)$ describes about the origin as s varies in K is deformed to the point $f_s(b) \neq 0$, and hence must cross the origin.)

Now note that $\{p_s(z)\}_{s \in K}$ is a family in $\mathrm{Rat}_{\mathbb{C}}(n)$ (notation as in Part VII, (2.1)(i)). Moreover $\pi_1(\mathrm{Rat}_{\mathbb{C}}(n)) \cong \mathbb{Z}$ (Part VII (2.5)) and is generated by a loop which moves a pole around a zero of a given $h \in \mathrm{Rat}_{\mathbb{C}}(n)$. Our example implies then that a generator of $\mathrm{Rat}_{\mathbb{C}}(n)$ gives a topological obstruction to a solution of the blending problem. This of course is not the only obstruction and in Section 6 we get an analytic obstruction based on the Schwarz lemma.

Stein Spaces (4.10). We would like to give now an interpretation of the blending problem as a (difficult) problem concerning Stein spaces. Recall that a complex space Ω (Gunning-Rossi [49]) is Stein if

(i) Ω may be covered by countably many compact subsets;

(ii) for each compact subset $S \subset \Omega$, $\hat{S} := \{z \in \Omega \mid |f(z)| \leq \sup_S |f|$ for all f holomorphic in $\Omega\}$ is compact;

(iii) if $z_1, z_2 \in \Omega$, $z_1 \neq z_2$, then there exists a holomorphic function f on Ω such that $f(z_1) \neq f(z_2)$;

(iv) if $\mathcal{O}_x :=$ ring of germs of holomorphic functions at $x \in \Omega$, and $m_x \subset \mathcal{O}_x$ is the maximal ideal, then there exist global holomorphic functions on which generate m_x for each $x \in \Omega$.

Now using the notation of (4.8) we set for each $s \in K$,

$\Gamma_s := \{(z, 1/p_s(z)) \mid z \in H\} \cap H \times \mathbb{C}$,

$\Gamma := \bigcup_{s \in K} \Gamma_s$,

$Y := H \times \mathbb{C} - \Gamma$.

Since the parametrization is continuous, and K is compact it is clear that Γ is closed in $H \times \mathbb{C}$, i.e. Y is an open subset of $H \times \mathbb{C}$.

Next consider the holomorphic map $\pi: Y \to H$ induced by the projection $H \times \mathbb{C} \to H$. Now recall from the discussion in (4.8) that to solve the blending problem we are required to find a meromorphic function $g(z)$ such that $1/g(z)$ is proper and holomorphic in $\bar{H}$, and $-g(z)+p_s(z) \neq 0$ for all $z \in \bar{H}$, $s \in K$. Let us weaken these requirements a bit, and consider the problem of finding a meromorphic function $\tilde{g}(z)$ such that $1/\tilde{g}(z)$ is holomorphic in H, and $-\tilde{g}(z)+p_s(z) \neq 0$ for all $z \in H$, $s \in K$. (Note that a solution to the blending problem satisfies these conditions automatically and it is a bit technically easier for us to work with the open half plane. Moreover such a solution corresponds to stabilizing the plant via dynamic output feedback, but not necessarily asymptotically stabilizing it.) Then the existence of such a function $\tilde{g}(z)$ is <u>equivalent</u> to the existence of a holomorphic section of $\pi: Y \to H$. Moreover if no such section exists, then the blending problem has no solution. But for the space Y we have:

<u>Proposition (4.11).</u> Y <u>is a Stein space</u>.

<u>Proof.</u> First $H \times \mathbb{C}$ is a Stein space (see e.g. [49]) and each Γ_s is one-dimensional. Thus by a theorem of Simha [135], $H \times \mathbb{C}-\Gamma_s$ is Stein for every $s \in K$.

Next in $\mathbb{C}^n$, a domain is Stein if and only if it is a domain of holomorphy (which may be defined for example by the property that there exists a holomorphic function which cannot be continued analytically beyond the domain; see Hormander [65], pages 36-44). But by [65], page 40, we have in general that if Ω_a is a domain of holomorphy for all a in some index set A, then the interior Ω of $\bigcap_A \Omega_a$ is a domain of holomorphy. In particular in our case $H \times \mathbb{C}-\Gamma = \bigcap_{s\in K} (H \times \mathbb{C}-\Gamma_s)$ and since each $H \times \mathbb{C}-\Gamma_s$ is Stein it is also a domain of holomorphy, so that $Y = H \times \mathbb{C}-\Gamma$ is a domain of holomorphy and hence Stein.

Q.E.D.

<u>Remark (4.12).</u> Even though Y is Stein, the requirement that $\pi: Y \to H$ has a holomorphic section is rather strong. For example, for $\pi: Y \to H$ to have a <u>continuous</u> section it is enough to require the fibers to be contractible. Noting that H is holomorphically equivalent to the open unit disc $D := \{z \in \mathbb{C} \mid |z| < 1\}$, one might be led to conjecture that for $W \subset D \times \mathbb{C}$ open Stein and $\pi': W \to D$ the holomorphic map induced by the natural projection $D \times \mathbb{C} \to D$, if the fibers of $\pi': W \to D$ are contractible, then π' admits a holomorphic section. This however is decidedly false:

<u>Example (4.13).</u> Set

$$W' := \{(z,w) \in \mathbb{C}^2 \mid |z|^2 + |w|^2 < 1\} \subset D \times \mathbb{C}$$

and let $\pi': W' \to D$ be induced by projection. Then we claim the only holomorphic section of π' is the trivial section $\sigma(z) = (z,0)$. Indeed, let $\sigma(z) = (z,f(z))$ be any holomorphic section. Then $|z|^2 + |f(z)|^2 < 1$. Now if f were non-constant, then f would take its maximum on the boundary of D which is impossible since

$|f(z)|^2 < 1-|z|^2$. Thus f must be constant and again from the inequality $|f(z)|^2 < 1-|z|^2$, we must have $f(z) = 0$ for all $z \in D$.

Now let $W = \{\mathrm{Re}\, w > 0\} \cap W'$. Then W is Stein (Gunning-Ross, [49]), but $\pi': W \to D$ will have no holomorphic section.

Remarks (4.14).

(i) Motivated by Montel's theorem [110], it is natural in trying to construct a holomorphic section for $\pi': W \to D$, to construct sections over compact subsets. One can construct however (Tannenbaum [147] (1.9)), examples in which the fibers of π' are bounded rectangles of fixed size (and so in particular contractible) such that $\pi': W \to D$ has no section over a compact subset of D .

(ii) However if we assume the boundary of W is sufficiently nice, one can construct examples where $\pi': W \to D$ does have a holomorphic section. For example, set

$$W := \{(z,w) \mid u(z) < \mathrm{Re}\, w < v(z)\}$$

where $u, -v$ are subharmonic, so that W is Stein (see Gunning-Ross [49]). Then there exist sections over compact subsets in D . If we assume that u, v are continuous on D and $u < v$ on the boundary of 0 , then one may show there will be sections over D (we exclude cases like $u \equiv 0$, $v(z) := 1-|z|$).

(iii) There are theorems in the literature (see e.g. Grauert [42], [43] and Grauert-Reckziegel [44]) which allow one to conclude positively in certain circumstances that holomorphic sections exist, and which allow one to solve the blending problem. Unfortunately, the hypotheses used in the theorems make the results not too interesting from a system theoretic standpoint. (An account is given in Tannenbaum [147].) Thus the blending problem in its full generality remains one of the important unsolved problems in system theory.

(iv) However, we can solve the blending problem in a very important special case (and perhaps the most important from the applied standpoint). This we do in Section 6. In order to give the solution we must first make contact with some classical results about interpolation with holomorphic functions. This will be done next in Section 5.

§5. Interpolation in the Unit Disc

We shall begin with some classical interpolation theory due to Nevanlinna, Pick and Schur among others. The basic modern reference for this is Walsh [149] (which has an extensive list of references) who bases his treatment on Nevanlinna [12].

Nevanlinna Interpolation (5.1). Nevanlinna interpolation deals with the problem of finding necessary and sufficient conditions for the existence of a holomorphic function $h: D \to D$ (D is the open unit disc) with the property that for $\alpha_1,\dots,\alpha_n \in D$, $\alpha_i \neq \alpha_j$ for $i \neq j$, and for $\beta_1^{(0)},\dots,\beta_n^{(0)} \in D$ (not necessarily distinct) we have $h(\alpha_k) = \beta_i^{(0)}$, $i = 1,\dots,n$. In (5.5) below we consider the case of interpolation with multiplicities.

Usually one requires that the solution $h(z)$ be such that $|h(z)| \leq M < 1$ for all $z \in D$. Since in Section 6 we will be interested in taking certain interpolated values arbitrarily close to the boundary of D instead of letting $M \to 1$ it will be more convenient just to allow $h: D \to \bar{D}$. Of course, if for some $\alpha \in D$, $|h(\alpha)| = 1$, by the maximum principle we would have that h is constant. In other words, if $h: D \to \bar{D}$ is not constant, then automatically we have $h: D \to D$.

We now proceed inductively to construct h (for complete details see Walsh [149], pages 286-289). So let $n = 1$. Then if $|\beta_1^{(0)}| = 1$, the function $h(z) \equiv \beta_1^{(0)}$ is the unique solution to the problem by the maximum principle. If $|\beta_1^{(0)}| < 1$, then $h(z) \equiv \beta_1^{(0)}$ works again, but so do infinitely many other non-constant functions. Indeed, let

$$(*) \qquad h_1(z) := \frac{h(z) - \beta_1^{(0)}}{1 - \overline{\beta_1^{(0)}}\, h(z)} \cdot \frac{1 - \overline{\alpha_1}\, z}{z - \alpha_1} .$$

Then since $|\beta_1^{(0)}| < 1$, if we assume $h: D \to D$ and $h(\alpha_1) = \beta_1^{(0)}$, then h_1 will also take the unit disc to itself. Moreover given any $h_1: D \to \bar{D}$, via (*) we may solve for h, and we will get a holomorphic function from the unit disc to itself such that $h(\alpha_1) = \beta_1^{(0)}$. Thus the interpolation problem has infinitely many solutions.

Now suppose $n > 1$. If $|\beta_1^{(0)}| = 1$, then the only possible holomorphic function $h: D \to \bar{D}$ satisfying $h(\alpha_1) = \beta_1^{(0)}$ must be $h(z) \equiv \beta_1^{(0)}$, and h will satisfy the other conditions $h(\alpha_i) = \beta_i^{(0)}$, $i = 2,\dots,n$ if and only if $\beta_1^{(0)} = \dots = \beta_n^{(0)}$. If $|\beta_1^{(0)}| < 1$, the existence and uniqueness of a solution h is equivalent by (*) above to the existence and uniqueness of a holomorphic function $h_1: D \to \bar{D}$ which has the property that if

$$(**) \qquad \beta_k^{(1)} := \frac{\beta_k^{(0)} - \beta_1^{(0)}}{1 - \overline{\beta_1^{(0)}}\, \beta_k^{(0)}} \cdot \frac{1 - \overline{\alpha_1}\, \alpha_k}{\alpha_k - \alpha_1} , \quad k = 2,\dots,n$$

then $h_1(\alpha_k) = \beta_k^1$, $k = 2,\dots,n$. If h_1 is any such holomorphic function, then (*) defines all possible solutions to our interpolation problem. We thus have the following theorem ([149], page 287):

<u>Theorem (5.2).</u> <u>A holomorphic function</u> $h: D \to \bar{D}$ <u>with the property that</u> $h(\alpha_i) = \beta_i^{(0)}$, $i = 1,\dots,n$, $\alpha_i \neq \alpha_j$ <u>for</u> $i \neq j$, $\alpha_i \in D$, $i = 1,\dots,n$ <u>will exist if and only if</u>

(a) <u>if</u> $|\beta_1^{(0)}| = 1$, <u>then</u> $\beta_1^{(0)} = \beta_2^{(0)} = \dots = \beta_n^{(0)}$;

(b) if $|\beta_1^{(0)}| < 1$, then h exists, if and only if there exists a holomorphic function $h_1 \colon D \to \bar{D}$ such that $h_1(\alpha_k) = \beta_k^{(1)}$, $k = 2,\ldots,n$ where the $\beta_k^{(1)}$ are as in (**).

Remark (5.3). The above argument gives an inductive procedure for the solution of the Nevanlinna interpolation problem. Indeed from (5.1) we have reduced the problem of finding an $h \colon D \to \bar{D}$ which takes given values at n points to that of finding an $h_1 \colon D \to \bar{D}$ which takes given values at $n-1$ points. So to complete the interpolation procedure, we play the same game again. Explicitly, let

$$(1) \qquad h_r(z) := \frac{h_{r-1}(z) - \beta_r^{(r-1)}}{1 - \beta_r^{(n-1)}\, h_{r-1}(z)} \cdot \frac{1 - \overline{\alpha_r}\, z}{z - \alpha_r}$$

$$h_0(z) := h(z)$$

$$(2) \qquad \beta_k^{(r)} := h_r(\alpha_k) = \frac{\beta_k^{(r-1)} - \beta_r^{(r-1)}}{1 - \overline{\beta_r^{(r-1)}}\, \beta_k^{(r-1)}} \cdot \frac{1 - \overline{\alpha_r} \cdot \alpha_k}{\alpha_k - \alpha_r}$$

for $k = r+1, r+2, \ldots, n$.

We thus have the following explicit solution to Nevanlinna interpolation ([149], pages 287-288):

Theorem (5.4). A holomorphic function $h \colon D \to \bar{D}$ with the property that $h(\alpha_i) = \beta_i^{(0)}$, $i = 1,\ldots,n$, $\alpha_i \neq \alpha_j$ for $i \neq j$, $\alpha_i \in D$ will exist if and only if we have one of the following two cases (notation as in (5.3)):

(a) $|\beta_1^{(0)}| < 1$, $|\beta_2^{(1)}| < 1$, $\ldots$, $|\beta_q^{(q-1)}| < 1$, $|\beta_{q+1}^{(q)}| = 1$,

$\beta_{q+1}^{(q)} = 1$, $\beta_{q+1}^{(q)} = {}^{(q)}_{q+2} = \ldots = \beta_n^{(q)}$;

(b) $|\beta_1^{(0)}| < 1$, $|\beta_2^{(1)}| < 1$, $\ldots$, $|\beta_n^{(n-1)}| < 1$.

In case (a), $h(z)$ is unique and given by (5.3) (1) and (2) where $h_q(z) \equiv \beta_{q+1}^{(q)}$. In case (b), $h(z)$ is not unique and all possible $h(z)$ are given by (5.3) (1) and (2), with $h_n(z)$ an arbitrary holomorphic function from the unit disc to itself.

Nevanlinna Interpolation with Multiplicities (5.5). We will need to consider in Section 6 also interpolations with multiplicities. We will do here the case of interpolations of multiplicity 2, leaving the reader to make the generalization to higher multiplicities (or see [149], pages 288-289).

We use the notation of (5.1) and (5.3). The problem is then, suppose we want to take α_1 of multiplicity 2 in h, i.e. we specify $h(\alpha_1) = \beta_1^{(0)}$, $h'(\alpha_1) = \beta_2^{(0)}$, where $h'(z)$ denotes the derivative of $h(z)$. But if $h(\alpha_1) = \beta_1^{(0)}$, we will have $h'(\alpha_1) = \beta_2^{(0)}$ for h defined by

$$(3) \qquad h_1(z) = \frac{h(z) - \beta_1^{(0)}}{1 - \overline{\beta_1^{(0)}}\, h(z)} \cdot \frac{1 - \overline{\alpha_1}\, z}{z - \alpha_1}$$

if and only if

$$(4) \qquad h_1(\alpha_1) = \beta_2^{(0)} \left(\frac{1 - |\alpha_1|^2}{1 - |\beta_1^{(0)}|^2} \right) =: \beta_2^{(1)}$$

(assuming the quotient exists).

Then h will exist if and only if

(a) $|\beta_1^{(0)}| = 1$, $\beta_2^{(1)} = 0$, or

(b) $|\beta_1^{(0)}| < 1$, $|\beta_2^{(1)}| = 1$, or

(c) $|\beta_1^{(0)}| < 1$, $|\beta_2^{(1)}| < 1$.

For (a) the only solution is $h(z) \equiv \beta_1^{(0)}$, and for (b) the only solution is given by (3) above, with $h_1(z) \equiv \beta_2^{(1)}$. For (c), $h(z)$ is given by (3) where h_1 is defined by

$$h_2(z) := \frac{h_1(z) - \beta_2^{(1)}}{1 - \overline{\beta_2^{(0)}}\, h_1(z)} \cdot \frac{1 - \overline{\alpha_1}\, z}{z - \alpha_1}$$

where h_2 is an arbitrary holomorphic function from the disc to itself. Thus we see there are no new methods involved for interpolations with multiplicities.

Nevanlinna-Pick Matrix (5.6). The kind of interpolation discussed above is actually very important in network theory and electrical engineering. For accounts of this see Youla-Saito [155] and Zeheb-Lempel [157].

In [155] (based on classical work of Nevanlinna [112] and Pick [114]) the authors consider the problem of interpolations $p: H \to H$ with p holomorphic and real. More specifically suppose we have a sequence $\alpha_1, \ldots, \alpha_n \in H$ of distinct points such that if $S := \{\alpha_1, \ldots, \alpha_n\}$ and $\alpha \in S$, then $\bar{\alpha} \in S$. Let $\beta_1, \ldots, \beta_n \in H$ (not necessarily distinct) such that if $S' = \{\beta_1, \ldots, \beta_n\}$ and $\beta \in S'$, then $\bar{\beta} \in S'$ and if α_i is real then β_i is real. Then we want to know when there exists a real, holomorphic function $p: H \to H$ (p is called a positive real function) such that $p(\alpha_i) = \beta_i$. Then it is shown ([155], page 87) that such a p exists if and only if the $n \times n$ Hermitian matrix (the Nevanlinna-Pick matrix)

$$Q := \left(\frac{\overline{\beta_r} + \beta_s}{\overline{\alpha_r} + \alpha_s} \right)_{r,s=1,\ldots,n}$$

is non-negative definite. (In the classical case [112], [114] as in (5.1), only the case $p: H \to H$ is considered without the requirement that p be real. Of course then we do not need the above conjugacy conditions on the α_i , β_i , and one derives an analogous matrix.)

Trivially since H is conformally equivalent to D, we may derive similar conditions for interpolations $D \to D$, $D \to H$, or $H \to D$. The latter case will be of greatest interest to us in Section 6. So let $c \in H$, and note that $\frac{z-c}{z+\bar{c}} : H \to D$. Given $\beta_i \in H$ as above, let $\gamma_i := \frac{\beta_i - c}{\beta_i + \bar{c}}$ for $i = 1,\ldots,n$. Then a simple calculation shows that the existence of a holomorphic function $\tilde{p}: H \to D$ such that $\tilde{p}(\alpha_i) = \gamma_i$, $i = 1,\ldots,n$ and such that the function $c(\frac{1+\tilde{p}(z)}{1+\tilde{p}(z)})$ is positive real, is equivalent to the condition that the $n \times n$ Hermitian matrix (also called the Nevanlinna-Pick matrix)

$$\tilde{Q} := \left(\frac{1 - \overline{\gamma_r \gamma_s}}{\overline{\alpha_r} + \alpha_s} \right)_{r,s=1,\ldots,n}$$

being non-negative definite.

§6. Feedback Stabilization of Plants with Uncertainty in the Gain Factor

We conclude these lecture notes with the promised solution of a special case of the blending problem (4.8). Throughout these notes, we have been discussing rather abstract methods from algebraic geometry, invariant theory, algebraic topology, etc. applied to system theory. Now system theory is after all an applied discipline, and so we want to illustrate how abstract mathematical methods (in this case interpolation theory) lead to specific system designs.

Let us review the discussion of (4.8). The blending problem concerns constructing an asymptotically stable controller Σ_2 and feedback sensor Σ_3 (with the assumptions of (4.8)) such that the closed loop defined by the diagram in (4.8) will also be asymptotically stable for a real scalar input/output plant Σ_1 with parameter uncertainty. In case Σ_1 has no parameter uncertainty, a complete solution is given in [154].

Now we want to consider a plant Σ_1 with the following kind of parameter uncertainty:

If $p(z)$ is the transfer function, we assume that $p(z) = k\tilde{p}(z)$ where $\tilde{p}(z)$ is a fixed real rational function and k the gain factor may vary in some real interval $[k_{min}, k_{max}]$. In other words for the family of rational functions $\{p_s(z)\}_{s \in K}$ of (4.8) we fix the roots and poles and only let the gain factor vary.

Let $c(z)$ be the transfer function for Σ_2 and $f(z)$ the transfer function for Σ_3 (all of our systems are scalar input/output defined over $\mathbb{R}$). Then the stability conditions we require amount to having $c(z)$ and $f(z)$ holomorphic in $\bar{H}$, and the denominator $1+p(z)c(z)f(z)$ of the closed loop transfer function should be strict Hurwitz i.e. have no zeros in $\bar{H}$.

This problem reduces to the following ([146]): Given a real proper rational function $\tilde{p}(z)$ (we take $\tilde{p}(z)$ to be proper instead of strictly proper since our argument below works also in this more general case and of course in particular implies the result for $\tilde{p}(z)$ strictly proper), let S be the set of all intervals $[a,b]$ $(b > a > 0)$ such that there exists $g(z)_{[a,b]}$ a strict Hurwitz rational function with real coefficients and $1/g(z)_{[a,b]}$ proper such that for all $k \in [a,b]$ $-g(z)_{[a,b]}+k\cdot\tilde{p}(z)$ is strict Hurwitz and $1-k\tilde{p}(\infty)/g(\infty)_{[a,b]} \neq 0$ (this last condition is put in for control theoretic reasons (see [154]); we do not regard it as crucial since it is automatically satisfied for the case of $\tilde{p}(z)$ being strictly proper and consequently has been ignored until now). Then the blending problem in this case amounts to finding an interval $[k_{min},k_{max}] \in S$ such that for all $[a,b] \in S$, $k_{max}/k_{min} \geq b/a$. Moreover we want to construct the corresponding rational functions $g(z)_{[k_{min},k_{max}]}$. This version of the blending problem we shall call the <u>blending problem with uncertainty in the gain factor.</u> To simplify the notation now on we drop the explicit reference to the dependence of $g(z)_{[a,b]}$ on $[a,b]$ and denote the function by $g(z)$.

We follow the treatment now of Tannenbaum [146]. We first recall the following definition from Youla <u>et al</u> [154], page 160:

<u>Definition (6.1).</u> Let $\tilde{p}(z)$ be as above. Denote the real zeros of $\tilde{p}(z)$ in $\tilde{H}$ $(=: \bar{H} \cup \{\infty\})$ by $z_1,\ldots,z_r$ and let the total number of real poles of $\tilde{p}(z)$ to the right of z_i (multiplicities included) be m_i, $i = 1,\ldots,r$. Then the roots and poles of $\tilde{p}(z)$ are said to have the <u>interlacing property</u> if all the m_i are odd or all the m_i are even.

We now have the following:

<u>Theorem (6.2). The following three conditions are equivalent:</u>

(a) <u>The roots and poles of</u> $\tilde{p}(z)$ <u>have the interlacing property.</u>

(b) <u>There exists a strict Hurwitz rational function</u> $g(z)$ <u>with real coefficients such that</u> $1/g(z)$ <u>is proper,</u> $-g(z)+\tilde{p}(z)$ <u>is strict Hurwitz, and</u> $1-\tilde{p}(\infty)/g(\infty) \neq 0$.

(c) <u>There exists a rational function</u> $h(z)$ <u>with real coefficients with the following properties:</u>

  (i) $h(z) \neq 1$ <u>for all</u> $s \in \tilde{H}$;

  (ii) <u>the zeros of</u> $h(z)$ <u>in</u> $\tilde{H}$ <u>are precisely the poles of</u> $\tilde{p}(z)$ <u>in</u> $\tilde{H}$, <u>multiplicities included</u>;

  (iii) <u>any zero of</u> $\tilde{p}(z)$ <u>in</u> $\tilde{H}$ <u>of multiplicity</u> m <u>is a pole of</u> $h(z)$ <u>of multiplicity at least</u> m.

<u>Proof.</u> The equivalence of (a) and (b) follows from [154], pages 161-164. For the equivalence of (b) and (c) just set $g(z) = \tilde{p}(z)h(z)$.

Q.E.D.

In these terms we have the following lemma:

<u>Lemma (6.3). The existence of a solution</u> $g(z)$ <u>to the blending problem with uncertainty in the gain factor which will work for all</u> k <u>contained in a maximal interval</u> $[k_{min},k_{max}]$ $(k_{max} > k_{min} > 0)$ <u>is equivalent to the existence of a</u>

rational function $h(z)$ with real coefficients with the following properties:

(i) $h(z) \notin [k_{min}, k_{max}]$ for all $z \in \tilde{H}$;

(ii) the zeros of $h(z)$ in $\tilde{H}$ are precisely the poles of $\tilde{p}(z)$ in $\tilde{H}$, multiplicities included;

(iii) any zero of $\tilde{p}(z)$ in $\tilde{H}$ of multiplicity m is a pole of $h(z)$ of multiplicity at least m .

Proof. As in (6.2), take $g(z) = \tilde{p}(z)h(z)$.

Q.E.D.

Remarks (6.4).

(i) From (6.2) if the roots and poles of $\tilde{p}(z)$ have the interlacing property, if we let the gain factor k vary on a sufficiently small neighborhood $U \ni 1$ on the real axis with $[a,b] \subset U$ and $1 \in [a,b]$, then by continuity one can always construct a corresponding solution $g(z)$ $(= g(z)_{[a,b]})$ for the blending problem with uncertainty in the gain factor. Thus the set S above is non-empty.

(ii) It is easy to see that the blending problem is trivial if $\tilde{p}(z)$ has no zeros or poles in H . Therefore, from (i) we make the following two assumptions about $\tilde{p}(z)$ which will hold for the rest of this section:

(a) $\tilde{p}(z)$ has at least one zero and at least one pole in H ;

(b) the roots and poles of $\tilde{p}(z)$ have the interlacing property.

Our first major problem is to prove in point of fact that the ratio k_{max}/k_{min} is finite. To do this we will first need to construct an explicit conformal equivalence ϕ from $\mathbb{P}^1-[a,b]$ to D $(\mathbb{P}^1: = \mathbb{C} \cup \{\infty\}$, $D: =$ open unit disc).

Construction of $\phi: \mathbb{P}^1-[a,b] \tilde{\to} D$ (6.5). We will construct a conformal equivalence $\phi: \mathbb{P}^1-[a,b] \to D$ $(b > a > 0)$ such that $\phi(0) = 0$. By the Schwarz lemma such an equivalence is essentially unique. We make this standard construction in several stages:

(1) Let $\phi_1: \mathbb{P}^1-[a,b] \to \mathbb{P}^1 - [-\infty,0]$ be defined by $\phi_1(z): = (z-a)/(z-b)$. Then $\phi_1(0) = a/b$.

(2) Let $\phi_2: \mathbb{P}^1 - [-\infty,0] \to H$ be defined by $\phi_2(z): = \sqrt{z}$. Then $\phi_2(a/b) = \sqrt{a/b}$.

(3) Let $\phi_3: H \to D$ be defined by $\phi_3(z): = \gamma(1-z)/(1+z)$ where $|\gamma| = 1$. Then

$$\phi_3(\sqrt{a/b}) = \gamma\left(\frac{1 - \sqrt{a/b}}{1 + \sqrt{a/b}}\right) = : \alpha' .$$

Note that $\alpha'/\gamma = |\alpha'|$ and $0 < |\alpha'| < 1$.

(4) Let $\phi_4: D \to D$ be defined by

$$\phi_4(z): = \gamma'(z-\alpha')/(\bar{\alpha}'z-1) \quad \text{where} \quad |\gamma'| = 1 .$$

Then $\phi_4(\alpha') = 0$.

Now set $\alpha: = \gamma'\alpha'$. Then defining $\phi: \mathbb{P}^1 - [a,b] \to D$ by $\phi: = \phi_4\circ\phi_3\circ\phi_2\circ\phi_1$, we have $\phi(0) = 0$, $\phi(\infty) = \alpha = \gamma'\gamma\left(\frac{1 - \sqrt{a/b}}{1 + \sqrt{a/b}}\right)$. Solving for b/a , we get that

$$b/a = \left(\frac{1+|\alpha|}{1-|\alpha|}\right)^2 .$$

Note as $|\alpha|$ approaches 1 , b/a approaches ∞ .

Theorem (6.6). Let $\tilde{p}(z)$ be as in (6.4)(ii). Then for any solution $g(z) = g(z)_{[k_{min},k_{max}]}$ to the blending problem with uncertainty in the gain factor $k_{max}/k_{min} < \infty$.

Proof. Let S be the set of all intervals $[a,b]$ $(b > a > 0)$ such that there exists $g(z)_{[a,b]}$ with the properties described above. Then we must show that $k_{max}/k_{min} := \max_{[a,b]\in S} b/a < \infty$. As in (6.3) for $[a,b] \in S$, to construct $g(z) = g(z)_{[a,b]}$, it is equivalent to constructing a real rational function $h: \tilde{H} \to \mathbb{P}^1 - [a,b]$ satisfying properties (ii) and (iii) of (6.3). Let $\psi: \bar{D} \to \tilde{H}$ be a conformal equivalence such that $\psi(0)$ = some zero of h in $\tilde{H}$ (the set of zeros of h in H are by (ii) required to be precisely the set of poles of $\tilde{p}$ in $\tilde{H}$). Then if $\phi: \mathbb{P}^1 - [a,b] \to D$, $\phi(0) = 0$ is as in (6.5), we have that the holomorphic mapping $\phi\circ h\circ\psi: \bar{D} \to D$ is such that $\phi\circ h\circ\psi(0) = 0$.

As in (6.5) set $\alpha_{[a,b]} := \phi(\infty)$ (we want to make the dependence of α on $[a,b]$ explicit now), and let z_0 be the pre-image under ψ of some pole of h which is also a zero of $\tilde{p}$ in H ((6.3)(iii)). Then $h(\psi(z_0)) = \infty$, and hence $\phi\circ h\circ\psi(z_0) = \alpha_{[a,b]}$. But by the Schwarz lemma $|z_0| \geq |\alpha_{[a,b]}|$ for all $[a,b] \in S$ (z_0 is independent of the intervals of S). Since

$$b/a = \left(\frac{1+|\alpha_{[a,b]}|}{1-|\alpha_{[a,b]}|}\right)^2$$

and $1 > |z_0|$, we see that $k_{max}/k_{min} < \infty$.

Q.E.D.

Remark (6.7). The fact that $k_{max}/k_{min} < \infty$ gives another obstruction to the blending problem (i.e. we cannot solve the problem for infinite variation of gain; see also [147]), this one essentially defined by the Schwarz lemma. Thus we have an analytic obstruction as compared to the topological obstruction of (4.9)

Solution of the Blending Problem (6.8). The proof of (6.6) contains the key idea for the solution of the blending problem with uncertainty in the gain factor. Explicitly from the properties (i)-(iii) of (6.3) we are required to construct a real rational function $h: \tilde{H} \to \mathbb{P}^1 - [a,b]$ with predetermined zeros (the poles of $\tilde{p}(z)$ in $\tilde{H}$) and with the set of poles containing a predetermined set (the zeros of $\tilde{p}(z)$ in $\tilde{H}$) and to find $b > a > 0$ such that b/a will be maximized relative to these properties (this is what we call k_{max}/k_{min}). Now since $\tilde{H}$ is conformally equivalent to $\bar{D}$ and $\mathbb{P}^1 - [a,b]$ to D , the problem of constructing a function with predetermined zeros and poles from $\tilde{H} \to \mathbb{P}^1 - [a,b]$ is equivalent to an interpolation problem from $\bar{D} \to D$. For the case of interpolations from $D \to D$ we have given the classical solution in Section 5. This method does not in general work for interpolations from $\bar{D} \to D$ and this has caused some problems in network theory which are discussed in Youla-Saito [155]. Thus now we will assume that all the finite roots and poles of $\tilde{p}(z)$ lying in $\bar{H}$ in point of fact lie in H. The case of roots and poles lying on the boundary is taken up in [146]. We will also assume for simplicity that the roots and poles of $\tilde{p}(z)$ are distinct. The interpolation method of Section 5 also works for multiple roots and poles as we have seen in (5.5).

Let $a_1,\ldots,a_n$ be all the poles of $\tilde{p}(z)$ lying in $\bar{H}$, and let $a_{n+1},\ldots,a_{n+m}$ be all the zeros of $\tilde{p}(z)$ lying in $\tilde{H}$. So we need $h: \tilde{H} \to \mathbb{P}^1 - [k_{min}, k_{max}]$ such that $h(a_i) = 0$, $i = 1,\ldots,n$, and $h(a_{n+j}) = \infty$, $j = 1,\ldots,m$. Let $\psi: \bar{D} \to \tilde{H}$ be defined by $\psi(z) := a_1(1+z)/(1-z)$. Note $\psi(0) = a_1$. Set

$$\alpha_k := \left(\frac{a_k}{a_1} - 1\right) / \left(\frac{a_k}{a_1} + 1\right) \quad \text{for } k = 1,\ldots,n+m .$$

Then $\alpha_1 = 0$, and $\psi(\alpha_k) = a_k$ for $k = 1,\ldots,n+m$. Let ϕ be as in (6.5) with $\phi(0) = 0$, $\phi(\infty) = \alpha$. Note that ϕ is determined only up to rotation, i.e. in Steps 3 and 4 of (6.5) we have two factors γ, γ' ($|\gamma| = |\gamma'| = 1$) coming into the definition of ϕ. However $|\alpha|$ is independent of γ, γ'.

Now via Nevanlinna interpolation (5.1) we can determine $\tilde{h}: D \to D$ such that $\tilde{h}(\alpha_i) = 0$, $i = 1,\ldots,n$ and $\tilde{h}(\alpha_{n+j}) = \alpha$, $j = 1,\ldots,m$. Note that an $\tilde{h}$ always exists for $|\alpha|$ sufficiently small by (6.2)(i) and the fact that we have assumed the roots and poles of $\tilde{p}$ have the interlacing property. Moreover we want to determine the maximum $|\alpha|$ such that $\tilde{h}$ will exist with these properties. It is easy to see that given such a solution $\tilde{h}$ we can always choose γ, γ' in the definition of ϕ to insure that $h := \phi^{-1} \circ \tilde{h} \circ \psi^{-1}$ will have real coefficients. Then from the fact that $b/a = [(1+|\alpha|)/(1-|\alpha|)]^2$, taking $|\alpha|$ maximum will give the desired value of k_{max}/k_{min}.

Note that since holomorphic functions take boundaries to boundaries, if we take an $\tilde{h}: D \to D$ corresponding to the maximum $|\alpha|$, the corresponding $g := \tilde{p}h$ will be strict Hurwitz, but $-g+k\tilde{p}$ will in general only be Hurwitz for $k \in [k_{min}, k_{max}]$, i.e. it may have zeros on the boundary of $\bar{H}$. See [146], page 10.

Finally we should remark that in our construction of $\psi: \bar{D} \to \tilde{H}$, the procedure we have given above to determine k_{max}/k_{min} is independent of the choice of pole $a \in H$ of $\tilde{p}(z)$ which we take such that $\psi(0) = a$. This follows immediately from the generalized form of the Schwarz lemma (Ahlfors [1], page 110, and Kobayashi [88], page 3) and the invariance of the Poincaré metric of the disc under conformal equivalence (Singer-Thorpe [136], pages 191-192).

An explicit example illustrating all these ideas is given in [146], pages 9-11.

Another Procedure for Computing k_{max}/k_{min} (6.9). Using the Nevanlinna-Pick matrix (5.6) one can also develop a procedure for computing k_{max}/k_{min}. Indeed let h be real rational with $h(z) \notin [a,b]$ for all $z \in \tilde{H}$ satisfying (ii) and (iii) of (6.3). Then $(h(z)-a)/(h(z)-b): \tilde{H} \to \mathbb{P}^1 - [-\infty,0]$ and so for all $z \in \tilde{H}$, the function

$$u(z) := +\sqrt{\frac{h(z)-a}{h(z)-b}}$$

must be positive real. In terms of $u(z)$ the conditions of (6.3) may be rephrased as:

(i') Re $u(z) > 0$, $u(z)^2$ is rational;

(ii') the zeros of $u(z) - \sqrt{a/b}$ in $\tilde{H}$ are the poles of $\tilde{p}(z)$ in $\tilde{H}$ multiplicities included;

(iii') every zero of $\tilde{p}(z)$ in $\tilde{H}$ is a zero of $u(z)-1$ of at least the same multiplicity.

Let $\ell(-z)$ be the polynomial whose zeros are the poles of $\tilde{p}(z)$ in $\tilde{H}$, multiplicities included. Then $\ell(z)$ is Hurwitz and

$$q(z) = \frac{\ell(-z)}{\ell(z)}$$

is analytic in $\tilde{H}$ and $q(z)q(-z) = 1$. Next set

$$w(z) = \frac{u(z) - \sqrt{a/b}}{u(z) + \sqrt{a/b}} .$$

Clearly $|w(z)| < 1$ for $z \in \tilde{H}$, and by (ii') and Schwarz's lemma $w(z) = q(z)w_1(z)$ for some bounded real function $w_1(z)$.

Obviously

$$w_1(z) = q(-z) \frac{u(z) - \sqrt{a/b}}{u(z) + \sqrt{a/b}}$$

and by (iii') every zero of $\tilde{p}(z)$ in $\tilde{H}$ must be a zero of

$$w_1(z)-q(-z) \quad \frac{1 - \sqrt{a/b}}{1 + \sqrt{a/b}}$$

of at least the same multiplicity. Moreover from (6.5)

$$|\alpha| = \frac{1 - \sqrt{a/b}}{1 + \sqrt{a/b}}$$

and we want to maximize $|\alpha|$.

We make the same assumptions now about the roots and poles of $\tilde{p}(z)$ as in (6.8). Let then the finite zeros of $\tilde{p}(z)$ in the upper quadrant of H and on the positive real axis be $s_1,\ldots,s_n$. From the above we must have that $w_1(z_i) = |\alpha|q(-s_i)$ for $i = 1,\ldots,n$. Consequently, if we let $q_i := q(-s_i)$, w_1 must take on the prescribed values $|\alpha|/q_i$ at the s_i . But by (5.6) a necessary and sufficient condition for the existence of such a w_1 is that the Nevanlinna-Pick matrix

$$\tilde{Q} = \left(\frac{1 - |\alpha|^2/\bar{q}_r q_t}{\bar{s}_r + s_t} \right)_{r,t=1,\ldots,n}$$

be non-negative definite.

Now define matrices

$$A: = \left(\frac{1}{\bar{q}_r + q_t} \right) \quad , \qquad B: = \left(\frac{\frac{1}{q_r q_t} - 1}{\bar{s}_r + s_t} \right) \quad .$$

It is easy to compute that $\tilde{Q}$ is non-negative definite if and only if

$$C: = \left(\frac{1 - |\alpha|^2}{|\alpha|^2} \right) A - B$$

is non-negative definite. Then it is simple to show ([155], page 104), that the lower bound on $(1 - |\alpha|^2)/|\alpha|^2$ such that C is non-negative definite is given by the largest eigenvalue λ_{max} of the pencil $\lambda A - B$. If we denote by $|\alpha|_{max}$, the maximum value of $|\alpha|$ corresponding to k_{max}/k_{min}, we have

$$\lambda_{max} = \frac{1 - |\alpha|^2_{max}}{|\alpha|^2_{max}}$$

and thus

$$k_{max}/k_{min} = \left(\frac{(1 + \sqrt{1+\lambda_{max}})^2}{\lambda_{max}} \right)^2 \quad .$$

The methods of (6.8) and (6.9) are explicitly compared in [146], pages 14-16.

BIBLIOGRAPHY

1. Ahlfors,L., Complex Analysis, Second Edition, McGraw-Hill, New York (1966).

2. Altman, A. and S. Kleiman, Introduction to Grothendieck Duality Theory, Lecture Notes in Math. 146, Springer-Verlag, Heidelberg (1970).

3. Antoulas, A.C., On canonical forms for linear constant systems, Report from the Mathematical System Theory Institute, E.T.H., Zürich, Switzerland (1978).

4. Antoulas, A.C., A polynomial matrix approach to F mod G-invariant subspaces, Ph.D. thesis, E.T.H., Zürich, Switzerland (1979).

5. Arnold, V.I., On matrices depending on parameters, Usp. Math. Nauk. 26, 101-114 (1971).

6. Atiyah, M.F. and I.G. MacDonald, Introduction to Commutative Algebra, Addison-Wesley Publ. Company, Reading, Massachusetts (1969).

7. Baras, J., R. Brockett and P. Fuhrmann, State space models for infinite dimensional systems, IEEE Trans. on Automatic Control AC-19, 693-700 (1974).

8. Birkhoff, G.D., A theorem on matrices of analytic functions, Math. Ann. 74, 122-133 (1913).

9. Birman, J.S., Braids, Links, and Mapping Class Groups, Annals of Math. Studies 82, Princeton University Press, Princeton, N.J. (1974).

10. Borel, A., Linear Algebraic Groups, Benjamin, New York (1965).

11. Bourbaki, N., Commutative Algebra, Addison-Wesley Puubl. Company, Reading, Massachusetts (1972).

12. Brockett, R., Some geometric questions in the theory of linear systems, IEEE Trans. on Automatic Control AC-21, 449-464 (1976).

13. Brockett, R., The geometry of the set of controllable systems, Res. Report of Aut. Contr. Lab., Nagoya University 24 (1977).

14. Brockett, R., The geometry of the partial realization problem, Proc. Conference on Decision and Control IEEE, New York (1978).

15. Brunovsky, P., A classification of linear controllable systems, Kibernetika 6, 176-188 (1970).

16. Bumby, R. and E. Sontag, Reachability does not imply coefficient assignability, Notices AMS (1978).

17. Byrnes, C., On the control of certain deterministic infinite dimensional systems by algebro-geometric techniques, Amer. J. Math. 100, 1333-1381 (1979).

18. Byrnes, C., Some geometric aspects of the output feedback problem, Lecture given at the NATO-AMS Adv. Study Institute and Summer Seminar on Algebraic and Geometric Methods in Linear System Theory, Harvard Univ., (June, 1979).

19. Byrnes, C. and P. Falb, Applications of algebraic geometry in system theory, Amer. J. Math. 101, 337-363 (1979).

20. Byrnes, C. and M. Gauger, Decideability criteria for the similarity problem, with applications to the moduli of linear dynamical systems, Adv. in Math. 25, 59-90 (1977).

21. Byrnes, C. and N. Hurt, On the moduli of linear dynamical systems, Adv. in Math. Supplementary Series, Studies in Analysis 4, 83-122 (1979).

22. Casti, J., Dynamical systems and their applications: linear theory, Math. in Science and Engineering 135, Academic Press, New York (1977).

23. Cauchy, A.L., Calculus des indices des fonctions, J. Ecole Polytech. 15, 176-229 (1937).

24. Chern, S.S., Complex Manifolds without Potential Theory, Van Nostrand Reinhold Company, New York (1967).

25. Clerk, J.M., The consistent selection of local coordinates in linear system identification, Proc. JACC, Purdue (1976).

26. Deligne, P., Equations Differentielles à Points Singuliers Réguliers, Lecture Notes in Math. 163, Springer-Verlag, Heidelberg (1970).

27. Dieudonne, J., Cours de Géométrie Algèbrique, Presses Universitaires de France, Paris (1974).

28. Eilenberg, S., Automata, Languages, and Machines, Vol. A, Academic Press, New York (1974).

29. Falb, P., Linear systems and invariants, Lecture Notes, Control Group, Lund University, Sweden (1974).

30. Fatou, P., Séries trigonométriques et séries de Taylor, Acta Math. 30, 335-400 (1906).

31. Fliess, M., Matrices de Hankel, J. Math. Pures Appl. 53, 197-224 (1974).

32. Fogarty, J., Invariant Theory, Benjamin, New York (1965).

33. Frobenius, G., Ueber Relationen zwischen den Näherungsbrüehen von Potensreihen, J. reine und angew. Math. 90, 1-17 (1881).

34. Frobenius, G., Ueber das Tragheitsgesetz quadratischen Formen, J. reine und angew. Math. 104, 187-230 (1895).

35. Fuhrmann, P., Algebraic system theory: An analyst's point of view, J. Franklin Inst. 301, 521-540 (1976).

36. Fuhrmann, P., On strict system equivalence and similarity, Int. J. Control 25, 5-10 (1977).

37. Fuhrmann, P., Linear feedback via polynomial models, Int. J. Control 30, 363-377 (1979).

38. Fuhrmann, P., Functional models, factorizations, and linear systems, Talks given at the NATO-AMS Adv. Study Inst. on Algebraic and Geometric Methods in Linear System Theory, Harvard Univ., (June, 1979).

39. Gantmacher, F.R., The Theory of Matrices, Vols. I and II, Chelsea, New York (1959).

40. Glover, K., Structural aspects of system identification, Ph.D. thesis, Dept. of Elect. Eng., M.I.T., Cambridge, Mass. (1973).

41. Grauert, H., Approximationssatze für holomorphe Funktionen mit Werten in komplexen Raumen, Math. Ann. 133, 139-159 (1957).

42. Grauert, H., Analytische Faserungen über holomorph-vollstandigen Raumen, Math. Ann. 135, 263-273 (1958).

43. Grauert, H. and H. Reckziegel, Hermitische Metriken und normale Familien holomorphen Abbildungen, Math. Zeitschrift 89, 108-125 (1965).

44. Griffiths, P. and J. Adams, Topics in Algebraic and Analytic Geometry, Mathematical Notes, Princeton Univ. Press, Princeton, N.J. (1974).

45. Griffiths, P. and J. Harris, Principles of Algebraic Geometry, Wiley, New York, (1978).

46. Grothendieck, A., Sur la classification des fibrés holomorphes sur la sphere de Riemann, Amer. J. Math. 79, 121-138 (1957).

47. Gunning, R., Lectures on Riemann Surfaces, Princeton Univ. Press, Princeton, N.J. (1966).

48. Gunning, R., Lectures on Vector Bundles over Riemann Surfaces, Princeton Univ. Press, Princeton, N.J. (1967).

49. Gunning, R. and H. Rossi, Analytic Functions of Several Complex Variables, Prentice-Hall, Englewood Cliffs, N.J. (1965).

50. Haboush, W.J., Reductive groups are geometrically reductive, Annals of Math. 102, 67-84 (1975).

51. Hanna, C., Decomposing algebraic vector bundles on the projective line, Proc. Amer. Math. Soc. 61, 196-200 (1976).

52. Hartshorne, R., Algebraic Geometry, GTM 52, Springer-Verlag, Heidelberg (1977).

53. Hazewinkel, M., Moduli and canonical forms for linear dynamical systems II: The topological case, Math. Systems Theory 10, 363-385 (1977).

54. Hazewinkel, M., Moduli and canonical forms for linear dynamical systems III: The algebraic geometric case, R. Hermann, C. Martin (editors), Proc. of the 1976 NASA-AMES Conf. on geometric control theory, Math. Sci. Press (1977).

55. Hazewinkel, M., On the (internal) symmetry groups of linear dynamical systems, P. Kramer, M. Dal-Cin (editors), Groups, systems and many-body physics, Vieweg (1979).

56. Hazewinkel, M., A partial survey of the uses of algebraic geometry in systems and control theory. To appear in Sym. Math. INDAM (Severi Centennial Conference, 1979), Academic Press.

57. Hazewinkel, M., (Fine) moduli (spaces) for linear systems: What are they and what are they good for, Lectures given at the NATO-AMS Study Inst. on Algebraic and Geometric Methods in Linear System Theory, Harvard Univ., (June, 1979).

58. Hazewinkel, M. and R. Kalman, On invariants, canonical forms and moduli for linear, constant, finite dimensional, dynamical systems, in Proc. CNR-CISM Symp. on Algebraic System Theory, Udine (1975), Lecture Notes in Economics Math. Syst. Theory 131, 48-60, Springer-Verlag, Heidelberg (1976).

59. Hazewinkel, H. and C. Martin, Symmetric groups, the specialization order, and systems, preprint (1980).

60. Hermann, R. and C. Martin, Applications of algebraic geometry to systems theory, Part I, IEEE Trans. on Automatic Control AC-22, 19-25 (1977).

61. Hermite, C., Sur le nombres de racines d'une equation algebrique comprise entre des limites donnes, J. reine und angewandte Math. 52, 39-51 (1856).

62. Heymann, M., Comments on "Pole assignment in multi-input controllable linear systems", IEEE Trans. on Automatic Control AC-13, 748-749 (1968).

63. Ho, B.L., An effective construction of realizations from input/output descriptions, Ph.D. thesis, Stanford University (1966).

64. Ho, B.L. and Kalman, R., Effective construction of linear state-variable models from input/output functions, Regelungstechnik 14, 545-548 (1966).

65. Hormander, L., Introduction to Complex Analysis in Several Variables, Van Nostrand, Princeton, N.J. (1966).

66. Horowitz, I., Synthesis of Feedback Systems, Academic Press, New York (1963).

67. Horowitz, I. and A. Gera, Blending of uncertain nonminimum-phase plants for elimination or reduction of nonminimum-phase property, Int. J. Systems Science 10, 1007-1024 (1979).

68. Horowitz, I. and U. Shaked, Superiority of transfer function over state variable methods in linear time-invariant feedback system designs, IEEE Trans. on Automatic Control AC-20, 84-97 (1975).

69. Horowitz, I. and M. Sidi, Optimum synthesis of nonminimum-phase feedback systems with parameter uncertainty, Int. J. Control 27, 361-386 (1978).

70. Hu, S.T., Homotopy Theory, Academic Press, New York and London (1959).

71. Humphreys, J., Linear Algebraic Groups, GTM 21, Springer-Verlag, Berlin and New York (1975).

72. Hurwitz, A., Über die Bedingungen unter welchen eine Gleichung nur Wurzeln mit negativen reelen Teilen besitz, Math. Ann. 52, 273-284 (1895).

73. Isidori, A. and A.J. Krener, Non-linear decoupling via feedback: A differential geometric approach, preprint (1979).

74. Jacobs, O.L., Introduction to Control Theory, Clarendon Press, Oxford (1974).

75. Kalman, R.E., Lectures on controllability and observability, Centro Internazionale Matematico Estivo Summer Course 1968, Cremonese, Rome.

76. Kalman, R.E., Pattern recognition properties of multilinear machines, IFAC Symposium, Yereyan, Armenian SSR (1968).

77. Kalman, R.E., On minimal partial realizations of an input/output map. In Aspects of Network and System Theory (edited by R. Kalman and N. DeClaris), Holt, Rinehart, and Winston, Inc., New York, 385-407 (1971).

78. Kalman, R.E., Kronecker invariants and feedback. In Ordinary Differential Equations (edited by L. Weiss), Academic Press, New York (1972).

79. Kalman, R.E., Algebraic geometric description of the class of linear systems of constant dimension, 8th Annual Princeton Conference on Information Sciences and Systems, Princeton, N.J., (March, 1974).

80. Kalman, R.E., System theoretic aspects of the theory of invariants, unpublished manuscript (1974).

81. Kalman, R.E., On partial realizations, transfer functions, and canonical forms, Acta Polytechnica Scandinavica 31, 9-32 (1979).

82. Kalman, R.E., M. Arbib and P. Falb, Topics in Mathematical System Theory, McGraw-Hill, New York, (1965).

83. Kamen, E., An operator theory of linear functional differential equations, J. Diff. Equations 27, 274-297 (1978).

84. Kimura, H., Pole assignment by gain output feedback, IEEE Trans. on Automatic Control AC-20, 509-516 (1975).

85. Kimura, H., A further result on the problem of pole assignment by output feedback, IEEE Trans. on Automatic Control AC-22, 458-463 (1977).

86. Kleiman, S., Geometry on Grassmannians and applications to splitting bundles and smoothing cycles, I.H.E.S. Pub. Math. 36 (1969).

87. Kleiman, S. and D. Laksov, Shubert calculus, Amer. Math. Monthly 79, 1061-1082 (1972).

88. Kobayashi, S., Hyperbolic Manifolds and Holomorphic Mappings, Marcel Dekker, New York (1970).

89. Kraft, H., Geometrische Methoden in der Invariantentheorie, Notes to a course given at the University of Bonn during the winter semester 1977-78.

90. Kraft, H. and C. Procesi, Closures of conjugacy classes of matrices are normal, Invent. Math. 53, 227-247 (1979).

91. Lang, S., Algebra, Addison-Wesley, Reading, Massachusetts (1971).

92. Langenhop, C., On the stabilization of linear systems, Proc. Amer. Math. Soc. 15, 735-742 (1964).

93. Luenberger, D.G., Introduction to Dynamic Systems, John Wiley and Sons, New York (1979).

94. Luna, D., Sur les orbites fermées des groupes algébriques réductifs, Invent. Math. 16, 1-5 (1972).

95. Lyapunov, A.M., Probleme général de les stabilité du mouvement, Ann. Fac. Sci. Toulouse 9, 203-474 (1907). (Reprinted in Ann. Math. Study No. 17, Princeton Univ. Press, Princeton, N.J. (1949).)

96. Mac Duffee, C.C., The Theory of Matrices, Chelsea, New York (1946).

97. Marden, M., The Geometry of the Zeros, Mathematical Surveys No. III, Amer. Math. Soc., New York (1949).

98. Martin, C. and R. Hermann, Applications of algebraic geometry to systems theory: The Mc Millan degree and Kronecker indices of transfer functions as topological and holomorphic invariants, SIAM J. Control and Optimization 16, 743-755 (1978).

99. Massey, W.S., Algebraic Topology: An Introduction, Harcourt, Brace and World, Inc., New York (1967).

100. Maxwell, J.C., On governors, Proc. Royal Soc. of London 16, 270-283 (1867/68).

101. Milnor, J., On the betti numbers of real varieties, Proc. Amer. Math. Soc. 15, 275-280 (1964).

102. Mislin, G., Finitely dominated nilpotent spaces, Ann. of Math. (2) 103, 547-556 (1976).

103. Morrow, J. and K. Kodaira, Complex Manifolds, Holt, Reinhart, and Winston, Inc., New York (1971).

104. Mumford, D., Geometric Invariant Theory, Ergeb. Math. Bol. 34, Springer-Verlag, Berlin and New York (1965).

105. Mumford, D., Lectures on Curves on an Algebraic Surface, Annals of Math. Studies 59, Princeton Univ. Press, Princeton, N.J. (1966).

106. Mumford, D., Introduction to Algebraic Geometry, Notes from Harvard Univ., Cambridge, Massachusetts.

107. Mumford, D., Algebraic geometry I: Complex projective varieties, Grundlehren der math. Weissenschaften 221, Springer-Verlag, Heidelberg (1976).

108. Mumford, D. and K. Suominen, Introduction to the theory of moduli, Proc. 5th Nordic Summer School in Math., Oslo, 1970 (edited by F. Oort), Wolters-Noordhoff, Groningen, 171-222 (1972).

109. Nagata, M., Lectures on the 14th problem of Hilbert, Tata Inst. of Fundamental Research Lecture Notes 31, Bombay (1965).

110. Narasimhan, Several Complex Variables, University of Chicago Press, Chicago and London (1971).

111. Nerode, A., Linear automaton transformations, Proc. Amer. Math. Soc. 9, 541-544 (1958).

112. Nevanlinna, R., Über beschrankte Funktionen, die in gegebenen Punkten vorgeschriebene Werte annehmen, Ann. Acad. Sci. Fenn. 13, No. 1 (1919).

113. Perron, O., Die Lehre von den Kettenbrüchen, Teubner, Leipzig (1913).

114. Pick, G., Über die Beschränkungen analytischer Funktionen, welche durch vorgegebenen Funktionswerte bewiskt sind, Math. Ann. 77, 7-23 (1916).

115. Popov, V.M., Invariant description of linear time-invariant controllable systems, SIAM J. Control 10, 252-264 (1972).

116. Quillen, D., Projective modules over polynomial rings, Inv. Math. 36, 167-171 (1976).

117. Richardson, R., Principal orbit types for algebraic transformation spaces in characteristic zero, Inv. Math. 16, 6-14 (1972).

118. Rosenbrock, H.H., State-space and Multivariable Theory, Nelson and Sons Ltd., London (1970).

119. Rouchaleau, Y., Linear, discrete time, finite dimensional dynamical systems over some classes of commutative rings, Ph.D. thesis, Stanford (1972).

120. Rouchaleau, Y. and E. Sontag, On the existence of minimal realizations of linear dynamical systems over Noetherian integral domains, J. Computer and System Sciences 18, 65-75 (1979).

121. Rouchaleau, Y. and B. Wyman, Linear dynamical systems over integral domains, J. Comput. Syst. Sci. 9, 129-142 (1975).

122. Rouchaleau, Y., B. Wyman and R. Kalman, Algebraic structure of linear dynamical systems III. Realization theory over a commutative ring, Proc. Nat. Acad. Sci. (USA) 69, 3404-3406 (1972).

123. Routh, E.J., Stability of a Given State of Motion, MacMillan, London (1877).

124. Segal, G., The topology of spaces of rational functions, Acta Mathematica 143, 39-72 (1979).

125. Serre, J.-P., Faisceaux algébriques cohérents, Ann. of Math. 61, 197-278 (1955).

126. Serre, J.-P., Géométrie algebrique et géométrie analytique, Ann. Inst. Fourier 6, 1-42 (1956).

127. Serre, J.-P., Algèbre Locale-Multiplicités, Lecture Notes in Math. 11, Springer-Verlag, Heidelberg (1965).

128. Serre, J.-P., Corps Locaux, Hermann, Paris (1968).

129. Seshadri, C.S., Triviality of vector bundles over the affine space K^2, Proc. Nat. Acad. Sci. U.S.A. 44, 456-458 (1958).

130. Seshadri, C.S., Mumford's conjecture for GL(2) and applications, in Algebraic Geometry (edited by S. Abhyankar) 347-371, Oxford University Press, London (1969).

131. Seshadri, C.S., Quotient spaces modulo-reductive algebraic groups, Ann. of Math. 95, 511-556 (1972).

132. Seshadri, C.S., Theory of moduli, Proc. A.M.S. Summer Inst. (Arcata), Amer. Math. Soc. Proc. Symp. Pure Math. 29, 263-304 (1975).

133. Silverman, L., Representation and realization of time-variable linear systems, Ph.D. thesis, Columbia Univ. (1966).

134. Silverman, L., Realization of linear dynamical systems, IEEE Trans. on Automatic Control AC-16, 554-567 (1971).

135. Simha, R., On the complement of a curve on a Stein space of dimension two, Math. Zeitschriff 82, 63-66 (1963).

136. Singer, I. and J. Thorpe, Lecture Notes on Elementary Topology and Geometry, Scott, Foresman and Co., Glenview, Illinois (1967).

137. Sontag, E., Linear systems over commutative rings: A survey, Ricerche di Automatica 7, 1-34 (1976).

138. Sontag, E., On split realizations of response maps over rings, Inf. and Control 37, 23-33 (1978).

139. Sontag, E., Polynomial response maps, Lecture Notes in Control and Information Sciences 13, Springer-Verlag, Heidelberg (1979).

140. Sontag, E., On the observability of polynomial systems I: Finite-time problems, SIAM J. Control and Optimization 17, 139-151 (1979).

141. Sontag, E. and Y. Rouchaleau, On discrete-time polynomial systems, J. Nonlinear Analysis, Methods. Theory, and Applications 1, 55-64 (1976).

142. Spanier, E.H., Algebraic Topology, McGraw-Hill, New York (1969).

143. Steenrod, N., The Topology of Fibre Bundles, Princeton Mathematical Series 14, Princeton University Press, Princeton, N.J. (1951).

144. Suslin, A., Projective modules over a polynomial ring, Dokl. Akad. Nauk. SSSR 26 (1976).

145. Sussmann, H., Existence and uniqueness of minimal realizations of nonlinear systems, Math. Systems Theory 10, 263-284 (1976/1977).

146. Tannenbaum, A., Feedback stabilization of linear dynamical plants with uncertainty in the gain factor, Int. J. Control 32, 1-16 (1980).

147. Tannenbaum, A., The blending problem and parameter uncertainty in control theory, preprint (1980).

148. Tannenbaum, A., Geometric invariants of linear systems, preprint (1980).

149. Walsh, J.L., Interpolation and approximation by rational functions in the complex domain, A.M.S. Colloquium Publications 20, Fourth Edition (1965).

150. Wang, S. and E. Davison, Canonical forms of linear multivariable systems, SIAM J. Control and Optimization 14, 236-250 (1976).

151. Wells, R.O., Differential Analysis on Complex Manifolds, Prentice-Hall, Inc., Englewood Cliffs, N.J. (1973).

152. Wonham, W.M., On pole assignment in multi-input controllable linear systems, IEEE Trans. on Automatic Control AC-12, 600-665 (1967).

153. Yeno, K. and S. Ishihara, Tangent and Cotangent Bundles, Marcel-Dekker, Inc., New York (1973).

154. Youla, D., J. Bongiorno and C. Lu, Single-loop feedback — stabilization of linear multivariable dynamic plants, Automatica 10, 159-173 (1974).

155. Youla, D. and M. Saito, Interpolation with positive-real functions, Journal of the Franklin Institute 284, 77-108 (1967).

156. Zariski, O. and P. Samuel, Commutative Algebra (Vols. I, II), Van Nostrand, Princeton, N.J. (1958),(1960).

157. Zeheb, E. and A. Lempel, Interpolation in the network sense, IEEE Trans. on Circuit Theory CT-13, 118-119 (1966).

INDEX

Absolutely minimal realization 93
Affine variety 1, 97
Algebraically canonical system 100
Algebraic group 36
    geometrically reductive 46
    linear 37
    linearly reductive 44
    radical of 45
    reductive 46
Algebraic observability 99
Associated sheaf 11

Blending problem 136
Braid group 110
Brunovsky canonical form 123

Canonical realization 33, 91
Canonical system 33, 91
Cauchy index 105
Causality 100
Čech cocycle 12
Coarse moduli space 39
Cocycle condition 12, 14
Coefficient assignability 130
Concatenation 19, 78
Connection 28
Constant sheaf 11
Control canonical form 27, 69
Controllability 22
Constructibility 24
Constructible set 9
Coordinate ring 6
Covariant differential operator 28
Covering space 109
Cyclic section 30

Differential equation of order n 29-30
    solution of 30
Dimension of variety 8
Divisor 113
Dual bundle 12
Dual system 93
Dynamical system 18
    constant 21
    continuous 20
    discrete 20
    finite dimensional 21
    in input/output or external sense 19
    linear 20
    smooth 20
    time-invariant 21
Dynamic output feedback 133

Equilibrium point 97, 119
asymptotically stable 120
stable 120
Event space 19

Family of deformations 67
Family of endomorphisms 38
Family of systems 55, 63, 89
Fatou ring 92
Feedback equivalence 126
Field of definition 16
Fine moduli space 39
Free system 26
Fuhrmann realization 86
Function field 8
Fundamental group 109

Gain factor 143
Gain matrix 133
Geometric quotient 44
universal 62
Graded ring 4
Grassmann variety 14

Hankel matrix 81, 91
Hermann-Martin bundle 125
Hermann-Martin map 124
Hilbert's 14th problem 45
Homogeneous coordinate ring 7
Homogeneous coordinates 4
Homogenized pencil 125
Homotopy 108
Horizontal section 28
Hurwitz polynomial 120

Identification theory 104
Impulse/response function 34
Impulse/response sequence 81
Input function 18
Input/output map 19
constant causal 100
discrete time constant linear 77
polynomial 101
Input values 18
Interlacing property 144
Irreducible 2

Jacobson radical 96
Jump point 89

Kalman realization 79
k-ideal 96
k-rational point 16
Kronecker indices 68
Kronecker nice selection 68
k-topology 17

Laplace transform 32
Left coprime factorization 85
Length of input sequence 77
Local holomorphic canonical form 67

Locally closed 9
Local moduli space 67
Local system 28
Loop space 115
Lyapunov function 121

McMillan degree 33, 107
Minimal partial realization 87
Minimal realization 80
Morphism 6, 97
    dominating 103
    projective 53
Mumford conjecture 46

Nerode equivalence 78
Nevanlinna interpolation 139-142
Nevanlinna-Pick matrix 142
Nice selection 52
Nilradical 96
n-th homotopy group 109

Observability 25, 91, 99
Observation algebra 99
Output feedback transformation 133
Output functions 18
Output values 18

Padé approximation 87
Partial Hankel matrix 88
Partial realization 76, 87
Pencil of matrices 125
Plücker map 15
Pole assignability 130
Polynomial response map 101
Polynomial system 97
Presheaf 10
Pre-stable point 52
Principal bundle 15
Projective space 3
Projective variety 4

Quasi-affine variety 2
Quasi-projective variety 4
Quasi-reachability 98
Quotient 43

Radical 2, 97
Rational action 44
Rational canonical form 27
Rational function 8
Reachability 22, 91, 98
Readout map 19
Realization 19, 33, 35, 77, 81, 84, 91, 102
Reduced k-algebra 11, 97
Regular map 6, 97
Relative homotopy group 110
Representable functor 39
Residue 86
Response map 101
Richardson's criterion 46

Riemann surface 28
Right coprime factorization 85
r-jet 29
Routh criterion 120

Scheme 3, 97
Section 10, 13
Serre conjecture 13
Sheaf 10
    locally free 11
    of modules 11
Shift operator 21, 77, 86
Signature 106
Split family 95
Split system 93
Stalk 10
State feedback 122
State module 90
State set 18
State space form 26
State transition function 19
Stein space 137
Strict equivalence 127
Structure sheaf 10
System over a ring 21, 90

Tangent space 12
Time set 18
Transfer function 33
    strictly proper 33, 84
Transition matrix 25

Unimodular 85
Universal bundle 60
Universal covering 110
Universal family 55
Universal quotient bundle 61
Universal subbundle 61

Vector bundle 11
    fiber of 12
    trivial 13
Versal morphism 66
Versal family of deformations 67
Volterra series 100

Zariski topology 1, 5, 97